Taxonomy of Finfish

Practical Manual

THE AUTHORS

N. Karthik is an Assistant Professor in the Department of Fisheries Resource Management at the College of Fisheries Science, St. Devasahayam Institute of Fisheries Science and Technology, Midalam, Kanniyakumari dist., Tamil Nadu. He graduated with a Bachelor's degree in Fisheries Science from Dr MGR. Fisheries College and Research Institute, Ponneri, Tamil Nadu. He had his Master's degree in Fisheries Science from the Faculty of Fisheries, Sher-E-Kashmir University of Agricultural Sciences and Technology. He has qualified for ASRB – NET (National Teaching Eligibility Test) in Fisheries Resource Management (2023) and cracked the ICAR-JRF examination (2021). During his professional tenure, he diligently collected over 100 different species from the entire Kanyakumari coast, preserving them at the "Mukkadal (Tri-seas) Marine Museum" in the College of Fisheries Science, DIFST. Additionally, he has successfully conducted a capacity-building programme with funding obtained from NFDB, Hyderabad. Currently, he is leading an institutional research project.

S. Shamini is an Assistant Professor in the Department of Fisheries Resource Management at the College of Fisheries Science, St. Devasahayam Institute of Fisheries Science and Technology, Midalam, Kanniyakumari dist, Tamil Nadu. She graduated with a Bachelor's degree in Fisheries Science from Dr MGR. Fisheries College and Research Institute, Ponneri, Tamilnadu and has a Master's degree in Department of Fisheries Biology and Resource Management at Fisheries College and Research Institute in Thoothukudi, Tamil Nadu. During her professional tenure, she has successfully conducted a capacity-building programme with funding obtained from NFDB, Hyderabad.

Prof. S. Felix is a distinguished figure in the field of fisheries education and technology consulting, with a career spanning over four decades. Currently serving as the Dean of the College of Fisheries Science at DIFST (Devasahayam Institute of Fisheries Science and Technology) in Midalam, Kanniyakumari district, Tamil Nadu and also President of AQUAFIN (Aquaculture Technology & Research Foundation India), he has held several prestigious positions, including Vice Chancellor of Tamil Nadu Fisheries University (TNJFU) and President of the World Aquaculture Society (WAS)-APC. Prof. Felix is recognized for his exemplary leadership and contributions to the fisheries and aquaculture sector, demonstrated through his involvement in various committees and organizations. He is a prolific author, having published numerous research papers, books, and manuals, and has played a pivotal role in introducing new fisheries science courses and establishing fisheries colleges and research centers.

Taxonomy of Finfish
Practical Manual

N. Karthik
S. Shamini
S. Felix

College of Fisheries Science
St. Devasahayam Institute of
Fisheries Science & Technology
Midalam, Kanniyakumari Dist.629193 (Tamil Nadu)

2025
Daya Publishing House®
A Division of
Astral International Pvt. Ltd.
New Delhi – 110 002

ISBN: 978-93-5919-712-8

Published by : **Daya Publishing House®**
A Division of
Astral International Pvt. Ltd.
– ISO 9001:2015 Certified Company –
4736/23, Ansari Road, Darya Ganj
New Delhi-110 002
Ph. 011-43549197, 23278134
E-mail: info@astralint.com
Website: www.astralint.com

Preface

Welcome to the Taxonomy of Finfish: Practical Manual. This manual is designed to assist individuals in the accurate identification and classification of finfish species. It covers essential techniques such as sample collection, preservation, morphometric measurements, and meristic counts. Additionally, it provides a systematic guide to identifying various finfish orders, along with advanced methods like DNA barcoding and protein analysis. We extend our sincere thanks to all contributors for their invaluable insights. We hope this manual serves as a practical tool for researchers, students, and professionals in the field of finfish taxonomy.

N. Karthik

S. Shamini

S. Felix

Contents

Chapter-1

Basics of Sample Collection and Preservation of Finfish

Introduction

Fisheries are one of the most important renewable resources. With increasing fishing pressure, the only option left for the sustainability of fisheries is their rational management. Proper management is possible with a thorough knowledge of the dynamics of the fish stocks. The role of taxonomy and proper identification cannot be overstressed in studies of population dynamics. Acquaintance with the main species should be such that there should no errors in identification of them in any special form such as racial differentiation, abnormalities, malformation due to decay or disease. As to species of less importance collections and observations can be made for taxonomic studies which will be useful in future. Species identification study is also a step towards understanding the bewildering biodiversity that characterizes in the marine ecosystem (**Joshi and Sreekumar, 2015**).

Aim

To know the basics of fish sample collection and preservation of finfish.

Sampling and Preservation

Sampling for Taxonomic Studies

For taxonomic study, specimens of all species occurring in the ecosystem covering the entire length range were collected and preserved in 5% formalin after injecting 5% formalin through the vent and dorsal musculature. The specimens were preserved in a wide mouthed bottle in such a manner that the shape is not distorted during storage.

Fish Collection Methods

The major objective of the bioinventory is to identify all the available species in the habitat using all the gear combinations. Two types of gears are employed can be divided in to two viz, active and passive categories. Passive gear is usually set and left stationary for a period and commonly used gear are gillnet and traps. Active gears used in the inventory are seine nets, trawl nets, dip nets, hooks and line and electric fishing. Different factors affect fish sampling such as water depth, conductivity, water clarity, water temperature, fish size and fish behaviour.

Fish Preservation Techniques

Careful and correct preservation procedures are important for getting good results of collected specimens. Usage of correct fixative, correct concentration, appropriate containers, clean and sharp dissecting tools, quality labels will enhance the quality and value of the preserved specimens. Voucher specimens are the representative samples of species identified in the field and preserved to verify the field identification. Type specimens are the name bearing specimen preserved to verify the taxonomy of species.

Preservation

All fish must be killed before the fixation and preservation. This can be achieved by keeping the fish in high doses of the anaesthetizing solution like clove oil; killing of fish is an ethical treatment of a live animal and also serves scientific purpose. Make a small incision on the right side of the abdomen to facilitate penetration of preservative. The right side is chosen for this and similar operations like removal of scale samples, because left side is used for getting morphometric data and is commonly shown in photographs.

Time of the formalin fixation of the average specimen may range from two days to a week. For permanent preservation formaldehyde can be removed by soaking the specimen at least for two days in fresh water with regular water change. The permanent preservation can be done in 70% of alcohol or 40 % alcohol.

Fixation

1. **Formalin:** It is commonly used preservative for specimens and it is available in the liquid form. About 10% of formalin is used for preservation of fish specimens. 10% solution of buffered formalin is prepared by combining in 1-part full strength formalin and 9 parts distilled water and add 3ml of borax per litre.
2. **Paraformaldehyde:** It is in a powder form and can be used to make formalin solution. A mixture of 1-part paraformaldehyde, 4 parts of

anhydrous sodium carbonate and a small amount of Alconox (wetting agent) can be added to 20g powder mixture of 400ml of distilled water to produce 10% buffered formalin solution.

3. **Alcohol:** Ethanol and Isopropanol are commonly used to fix and preserve fish specimens.

Fixation Procedure

Specimen should be fixed immediately after collection. To fix the specimen wide mouthed glass filled with fixative solution is can be used. Tight lids must be used for fixation. The specimen must be preserved in as natural as possible. The fixative should be injected to the body cavity to facilitate penetration and preservation of internal organs.

Labelling

Labelling has to be done neatly and should not make any confusion in the future. Labels include date of collection, place of collection, ecological details, method of capture and name of collector.

INFERENCE:

Chapter-2

General Procedure for Species Identification of Finfish

Introduction

To learn the characters of importance for the identification of fishes and by which they may be accurately identified. Line drawings, colour plates and photographs provide basis for the learning salient characters used for their classification. Identification keys can be used as distinguishing characters of each family and order in which forms are treated according to phylogeny.

Aim

To know about the general procedure for species identification of finfish.

Procedure

- Determine the family in "key" to the families.
- Identify the lowest taxonomic unit listed in key to the family of which the fish is member.
- Verify the final determination by ascertaining by comparing the similarities of the specimen with illustration.
- The specimen collected matches specimens previously identified by taxonomists.
- Confirm the geographical range as given in the standard texts includes the locality from which the specimen was taken.

- Compare the description given in the FAO identification sheets, Smitha sea fishes, Catalog of Fishes and Fish base.

INFERENCE:

Chapter-3

Preservation of Fish by Using Formalin

Aim

The aim of this practical exercise is to preserve the specimen for taxonomic study.

Materials Required

- Fresh specimen
- Glass containers with tight-sealing lids
- 90% Distilled water and 10% Formalin
- Dissection tools (scissors, forceps etc.)
- Labels
- Permanent markers
- Data recording sheets
- Refrigerator

Procedure

Preparation

- Ensure that all materials and equipment are clean and sterile.
- Set up a clean and organized workspace for the preservation process.

Collection and Identification

- Obtain fresh specimens from a reliable source.
- Confirm the taxonomic identity of the specimens by examining external features and using field guides or taxonomic keys.

Labelling

- Prepare labels with the following information:
- Date of collection
- Location of collection
- Common name
- Scientific name
- Length of the specimen
- Weight of the specimen
- Method of capture
- Name of collector
- Attach labels to the glass containers.

Preservation

- Fill the glass containers with enough 90% Distilled water and 10% Formalin to completely submerge the specimens.
- Gently place the specimens into the Formalin-filled containers.
- Ensure that the specimens are fully submerged and free from air bubbles. You may use forceps to position them properly.
- Seal the containers tightly with their lids to prevent any evaporation or contamination.

Data Recording

- Record additional information about the collected specimens, such as:
- Size of the specimens
- Weight of the specimens
- Any peculiar morphological features
- Record this information on the data recording sheets.

Storage

- Store the sealed glass containers with the preserved specimens in a cool and dark place. A refrigerator is ideal for long-term storage.
- Make sure to maintain a constant temperature to avoid fluctuations.

Clean-up

- Dispose of any waste materials properly and clean your workspace thoroughly.

Data Management

- ❑ Organize and store all collected data and labels in a safe and easily accessible location.

INFERENCE:

Chapter-4

Morphometric Measurements Procedure of Fish

Introduction

Morphometric measurements of fish species are essential for understanding their biology, ecology, and population dynamics. Conducting a practical experiment to obtain these measurements can be a valuable exercise for fisheries scientists and students.

Aim

The aim of this practical exercise is to measure the morphometric features of fishes for taxonomical study.

Materials and Equipment

Specimens: Obtain fresh or preserved specimens.

Measuring Instruments

- Vernier calipers
- Measuring tape
- Scales (preferably digital)
- Slotted weights
- Dissecting kit (scissors, forceps, and a scalpel)

Lab Equipment

- ❑ Dissection trays
- ❑ Glass containers
- ❑ Ethanol for preservation (if working with fresh specimens)
- ❑ Camera (optional for photographic documentation)

Data Recording Tools

- ❑ Data sheets or notebooks
- ❑ Pencils and waterproof pens
- ❑ Ruler or graph paper for drawings

Procedure

Specimen Preparation

- ❑ If using fresh specimens, measure and record the total length (TL) as soon as possible. TL is the length from the tip of the snout to the end of the tail.
- ❑ Optionally, photograph each specimen to maintain a visual record.

Morphometric Measurements

- ❑ Lay the fish on a flat surface with its ventral side facing up.
- ❑ Use vernier calipers to measure various morphometric characteristics, including:
- ❑ **Total Length (TL):** The greatest dimension between the most anteriorly projecting part of the head and the farthest tip of the longer lobe of caudal fin when the caudal rays are spread out together.
- ❑ **Standard Length (SL):** The distance from the anterior most part of the head backward to the end of the vertebral column (structural base of caudal rays).
- ❑ **Fork Length (FL):** Distance from the tip of snout to the end of the middle ray of the caudal fork when the fish is being flattened out.
- ❑ **Head Length (HL):** Taken from the tip of the snout to the posterior most point reached by the bony margin of the operculum.
- ❑ **Pre-orbital length (PROL):** Distance from the tip of the snout to the forward point of eye.
- ❑ **Eye diameter (ED):** Horizontal diameter of the visible part of the eye, *i.e.,* the distance between the front edge and the back edge of the orbit.
- ❑ **Postorbital length (PSOL):** Distance from the backward point of eye to middle of the backward bony edge of the operculum.
- ❑ **Upper jaw length (UPJL):** Length of maxillary is taken from the anterior most point of the premaxillary to the posterior point of the maxilla.

- ❑ **Lower jaw length (LOJL):** Length of lower jaw from anterior tip to angle of mouth.
- ❑ **Body depth (BDD):** Distance between the middle point of dorsal finbase to straight downward central margin of the body, excluding fins.
- ❑ **Pre-dorsal length 1 (PD1):** Distance from the tip of the snout to the forward origin of the dorsal (intersection point of the forward edge of the first ray of the dorsal, D1, with the outline of the back, the fish being flattened out)
- ❑ **Pre-dorsal length (PD2):** Distance from the tip of snout to the forward origin of the dorsal (intersection point of the forward edge of the first ray of the dorsal, D2 , with the outline of the back, the fish being **flattened out).**
- ❑ **Pectoral fin length (PEL):** Distance from the extreme base of the uppermost ray to the farthest tip of the fin, filament if any.
- ❑ **Pelvic fin length (PVL):** Distance from the extreme base of the uppermost ray to the farthest tip of the fin, filament if any.
- ❑ **Dorsal fin length 1 (DF1):** Distance from the origin of the tip of the fin to the anterior lobe.
- ❑ **Dorsal fin length 2 (DF2):** Distance from the origin of the tip of the fin to the anterior lobe.
- ❑ **Inter dorsal length (IDL):** Distance from the base of the last spine (ray) of first dorsal to the intersection point of second dorsal fin.
- ❑ **Pectoral fin base length (PEB)**: Distance from the base of the anterior fin ray of the pectoral (P) to the backward end of the last ray, the pectoral being extended on the side of the fish in its normal position.
- ❑ **Pelvic fin base length (PVB):** Distance from the base of the anterior fin ray of the pelvic fin (P) to the backward end of the last ray, the ray being extended on the side of the fish in its normal position.
- ❑ **Dorsal fin base length (DB1):** Distance from the forward origin of the dorsal (D1) to the backward edge (Intersection point of the backward edge of the last spine, D', with the outline of the back, the fin being extended).
- ❑ **Dorsal fin base length (DB2):** Distance from the forward origin of the dorsal (D2) to the backward edge (Intersection point of the backward edge of the last ray, D2, with the outline of the back, the fin being extended.
- ❑ **Anal fin length (AFL):** Distance from the origin of the tip of the fin to the anterior most outer tip of the anal fin.
- ❑ **Anal fin base length (ABL):** Distance from the forward origin of the anal (A) to its backward edge (intersection point of the backward edge of the last ray, A' with outline of the abdomen, the fin being extended).
- ❑ **Caudal peduncle length (CPL):** Distance from the base of the second dorsal end to origin of the caudal fin.

- **Caudal peduncle depth (CPD):** Depth of the caudal peduncle.
- **Pre-pelvic length (PRP):** Distance from the tip of the snout to the anterior origin of the pelvic (intersection point of the forward edge of the first ray of the pelvic, with the contour of the abdomen, the fin being extended).
- **Pre-pectoral length (PRV):** Distance from the tip of the snout to the margin of the insertion of pectoral fin.
- **Pre-anal distance (PRA):** Distance from the tip of the snout to the forward origin of the anal (interior point of the forward edge of the first ray of the anal, A, with the outline of the abdomen, the fin being extended).
- Record measurements in millimeters for each fish specimen on your data sheet.

Weight Measurement

- Weigh each fish specimen using a weighing balance. Record the weight in grams (g).
- To determine the fish's condition factor (K), calculate K = (Weight / SL^3) * 100.

$$K = \frac{\text{Weight}}{SL^3} \times 100$$

Internal Examination (Optional):

- If you have the appropriate facilities and expertise, you can dissect a few specimens to examine internal morphometric features, such as gonad development, stomach contents, and otoliths.

Data Analysis:

- Analyze the collected data to calculate mean, standard deviation, and other relevant statistical measures.
- Compare the morphometric data among the sampled fish to identify any size or shape variations.

Results Presentation:

- Create graphs, tables, and diagrams to present your findings.
- Write a report summarizing your experiment, including the morphometric data, observations, and any significant findings.

INFERENCE:

Chapter-5

Meristic Counts Procedure of Finfish

Definition

Meristic features are countable structures of fishes such as fin spines and rays, gill rakers, lateral line scales, and branchiostegal rays.

Aim

The aim of this practical exercise is to count the meristic features of the finfish for taxonomic study.

Materials and Equipment

Specimen: Obtain fresh or preserved specimens.

Counting Instruments:

- Magnifying glass or dissecting microscope
- Fine-tip forceps
- Ruler
- Data recording tools (notebook, waterproof pens)

Lab Equipments:

- Dissection trays
- Glass containers
- Ethanol for preservation (if working with fresh specimens)

Procedure

Specimen Preparation

- ❑ Measure and record the total length (TL) of the fresh specimen. TL is the length from the tip of the snout to the end of the tail.
- ❑ Maintain a visual record of the specimen by using a photograph.

Meristic Measurements:

- ❑ Place a single specimen in a dissection tray with its ventral side facing up.
- ❑ Using a magnifying glass to visualize and count meristic features
- ❑ Count and record the following meristic characters for each specimen
- ❑ **Dorsal fin spines**: Count and record the number of spines in the first dorsal fin and the second dorsal fin.
- ❑ **Anal fin spines**: Count and record the number of spines in the anal fin.
- ❑ **Pelvic fin rays**: Count and record the number of rays in the pelvic fins.
- ❑ **Pectoral fin rays**: Count and record the number of rays in the pectoral fins.
- ❑ **Adipose fin rays:** Count and record the number of rays in the Adipose fins.
- ❑ **Caudal fin rays**: Count and record the number of rays in the caudal fin.
- ❑ **Lateral line scales**: Count and record the number of scales along the lateral line.
- ❑ **Gill rakers**: Count and record the number of gill rakers on one gill arch.

Data Recording

- ❑ Record all meristic measurements given specimens.
- ❑ Ensure that each measurement corresponds to the specific specimen it was obtained from.

Data Analysis

- ❑ By analyzing the meristic data to calculate the mean, range, and variation in the meristic characters among the sample fish.

Results Presentation

- ❑ Create tables to present your findings, summarize the meristic measurements for the specimen.
- ❑ Write a report detailing the meristic characteristics and variations observed.

Comparative Analysis

- ❑ Compare your meristic measurements with established data to aid in species identification and classification.

INFERENCE:

Chapter-6

Identification of Order: Hexanchiformes

Introduction

Hexanchiformes is an order comprising seven extinct species, representing the most primitive types of sharks. Fossil evidence suggests similarities to modern seven gill sharks dating back to the Jurassic period.

Aim

The aim of a practical exercise is focused on the identification of the order Hexanchiformes.

Systematic Classification

Kingdom: Animalia

Phylum: Chordata

Class: Chondrichthyes

Order: Hexanchiformes

Family: Hexanchidae

Identification Key Characters of Family – Hexanchidae (Cowsharks, Sixgill, and Sevengill Sharks)

- **Size and Body:** Small to large sharks with varying body shapes, not eel shaped.
- **Gill Slits:** Head with 6 or 7 pairs of long gill slits, last in front of pectoral-

fin origins, first pair not connected across the throat.

- **Eyes:** No nictitating lower eyelids; short, pointed snout; nostrils without barbels or nasoral grooves.
- **Mouth:** Very long mouth extending far behind the eyes; teeth of upper and lower jaws differ, with saw-like lowers having large cusps or cusplets.
- **Gill Structure:** Short dermal gill rakers present on inner gill slits; spiracles present and small.
- **Fins:** Single dorsal fin posterior to pelvic fins; anal fin present; caudal fin less than half total length, strongly asymmetrical with a subterminal notch and a very short lower lobe.
- **Caudal Peduncle:** Not depressed, without keels; no precaudal pits.

Layout

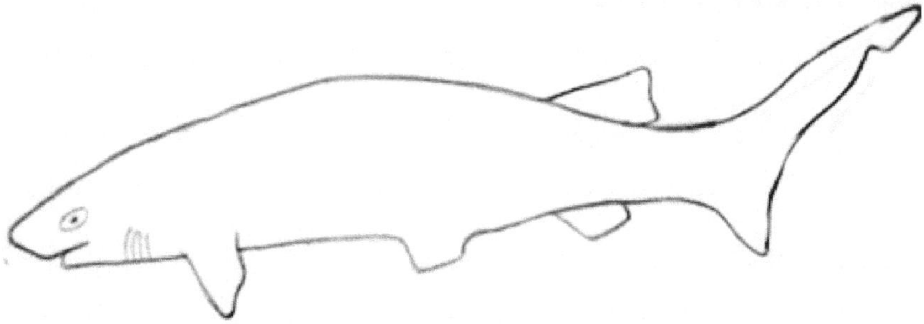

INFERENCE:

Chapter-7

Identification of Order: Squaliformes

Introduction

Squaliformes, a diverse order encompassing approximately 126 species distributed across seven families, exhibit distinct characteristics. These sharks typically feature two dorsal fins, often accompanied by spines, and possess sharp-headed profiles. They lack an anal fin or nictitating membrane and typically have five to seven gill slits. Varied in both form and size, Squaliformes are commonly found in saltwater or brackish habitats, ranging from shallow coastal seas to the open ocean. Their distribution spans global waters, from northern regions to tropical zones.

Aim

The aim of a practical exercise is focused on the identification of the order: squaliformes

Systematic Classification

Kingdom: Animalia

Phylum: Chordata

Class: Chondrichthyes

Order: Squaliformes

Family: Echinorhinidae and Squalidae

Identification Key Characters of Family – Echinorhinidae (Bramble sharks)

- **Appearance:** Stout cylindrical bodies without abdominal ridges; moderately depressed head.
- **Gill Slits:** Fifth gill slits abruptly expanded; very small spiracles well behind eyes; far-apart nostrils.
- **Mouth:** Short snout; broad, arched mouth; very short labial furrows without encircling; non-papillose lips.
- **Teeth:** Alike in both jaws; strongly compressed, blade-like, with a cusp and up to three side cusplets in adults.
- **Fins:** Two small spineless dorsal fins close together, both smaller than pelvic fins; origin of first dorsal fin behind

Layout

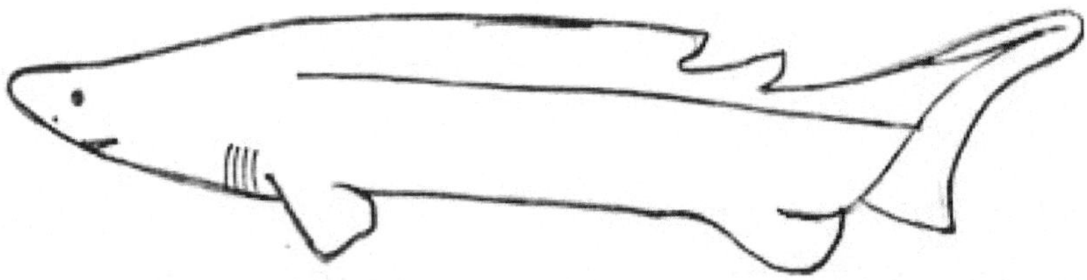

Identification Key Characters of Family – Squalidae (Dog fish sharks)

- **Appearance:** Small to moderately large, cylindrical bodies; no ridges between pectoral and pelvic fins; caudal keels and usually precaudal pits on the caudal peduncle.
- **Gill Slits:** Head with 5 gill slits, all anterior to pectoral fins; fifth not abruptly longer; large spiracles just behind eyes.
- **Snout and Mouth:** Short to moderately elongated, broad, flattened snout; no rostral saw; wide-spaced nostrils; short, nearly transverse mouth with smooth lips.
- **Teeth:** Strong-cusped, similar in both jaws, compressed, broad, blade-like, without cusplets; upper teeth nearly as large as lowers.
- **Fins:** Two dorsal fins with a long, strong, ungrooved spine on anterior margins; first dorsal fin larger or subequal to the second; origin far in front of pelvic fins; no anal fin; strongly asymmetrical caudal fin with subterminal notch.

Layout

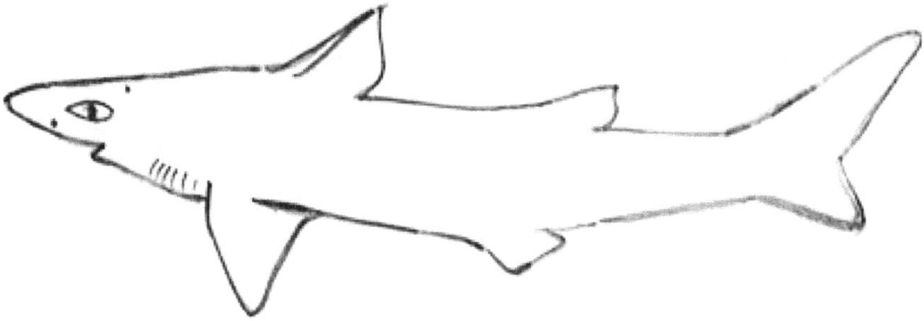

INFERENCE:

Chapter-8

Identification of Order–Orectolobiformes

Introduction

Orectolobiformes, encompassing moderate-sized to gigantic sharks, exhibit varied behaviors ranging from sluggish and demersal to active. These sharks lack nictitating eyelids and feature distinctive facial characteristics, including barbels and nasoral grooves. Their short mouths terminate in front of the eyes, with no enlarged anterior teeth or small intermediate teeth between the anteriors and lateral teeth on each side of the upper jaw. They possess five narrow to broad gill openings and are equipped with two spineless dorsal fins and an anal fin.

Aim

The aim of a practical exercise is focused on the identification of the order: Orectolobiformes.

Systematic Classification

Kingdom: Animalia

Phylum: Chordata

Class: Chondrichthyes

Order: Orectolobiformes

Family: Hemiscyllidae, Stegostomatidae, Ginglymostomatidae and Rhincodontidae

Identification Key Characters of Family – Hemiscyllidae (Bamboo sharks)

- **Spiracles:** Large.
- **Anal Fin Origin:** Well behind the second dorsal-fin origin.
- **Spacing:** Separated from lower caudal-fin origin by a space less than its base length.
- **Nasal Barbels:** Short.
- **Chin Groove:** Absent.
- **Anal Fin:** Low, rounded, and keel-like.
- **Body Proportions:** Distance from vent to lower caudal-fin origin longer than the distance from snout to vent.

Layout

Identification Key Characters of Family – Stegostomatidae (Zebra Shark)

- **Body:** Cylindrical, slightly flattened head, short blunt snout.
- **Gill Slits:** Five short gill slits, last three over pectoral fin bases, fourth and fifth closer together.
- **Nostrils:** Short barbel and groove from each nostril to the mouth.
- **Mouth:** Nearly straight, three lobes on lower lip, furrows at corners.
- **Teeth:** 28–33 tooth rows in upper jaw, 22–32 in lower jaw, with a large central cusp flanked by two smaller ones.
- **Ridges:** Five distinctive ridges in adults, one along dorsal midline, two on sides.
- **Fins:** First dorsal fin halfway along body, twice the size of the second dorsal fin. Large pectoral fins, smaller pelvic and anal fins.
- **Caudal Fin:** Almost as long as body, barely developed lower lobe, strong ventral notch near upper lobe tip.
- **Size:** Up to 2.5 m (8.2 ft), with rare reports of 3.5 m (11 ft).

- **Rare Morph:** Sandy zebra shark, overall sandy-brown, lacking distinct dark-spotted pattern.

Layout

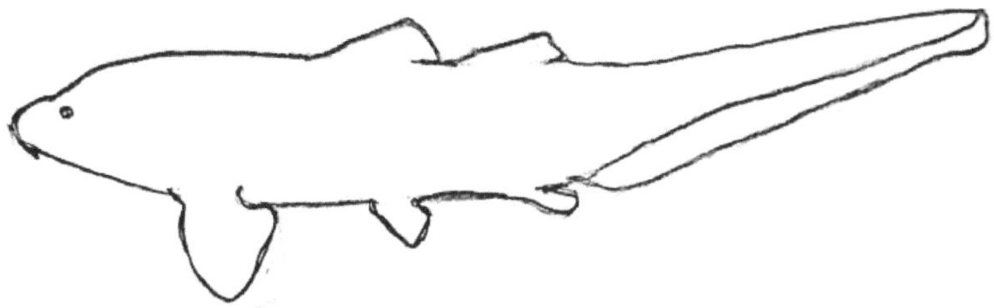

Identification Key Characters of Family – Ginglymostomatidae (Nurse Sharks)

- **Color:** Yellowish to dark brown.
- **Fins:** Muscular pectoral fins, two spineless dorsal fins (second smaller) aligned with pelvic and anal fins, tail exceeding one quarter of body length.
- **Mouth:** Highly distinctive, subterminal (far ahead of eyes and before snout), reflecting bottom-dwelling nature.
- **Barbels:** Two fleshy barbels on the lower jaw, aiding in prey detection in sediments.
- **Nostrils:** No circumnarial lobe and groove around outer edges of nostrils
- **Eyes:** Mouth far ahead of eyes, emphasizing their benthic (bottom-dwelling) behavior.
- **Spiracles:** Small, circular openings behind each eye as part of the respiratory system.
- **Teeth:** Serrated, fan-shaped, continuously replaced throughout the shark's life.

Layout

Identification Key Characters of Family – Rhincodontidae (Whale Shark)

- **Mouth:** Huge and nearly terminal (at the end of the snout).
- **Gill Slits:** External gill slits are very large, and there are internal gill slits inside the mouth cavity with filter screens.
- **Caudal Peduncle:** Strong lateral keels.
- **Caudal Fin:** Strong ventral lobe, lacking a strong terminal lobe and subterminal notch.

Layout

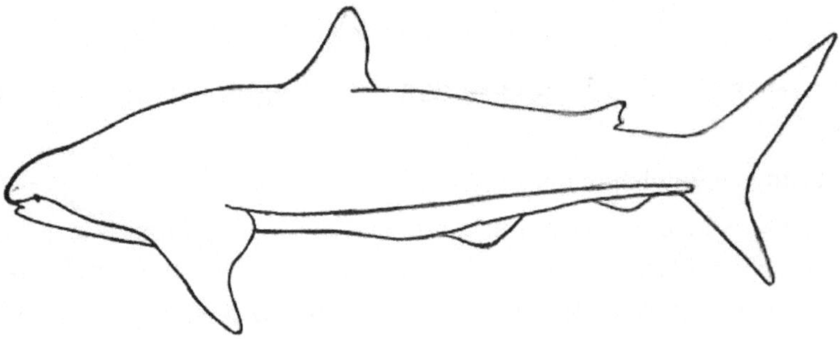

INFERENCE

Chapter-9

Identification of Order – Lamniformes

Introduction

Belonging to the Lamniformes order, these creatures typically possess a fusiform body shape tailored for swift swimming, accompanied by five gill slits and sharp, serrated teeth. They commonly feature two dorsal fins, each adorned with a spine, and a heterocercal tail with the upper lobe larger. Typically, they lack an anal fin but have a well-developed nictitating membrane over their eyes. Their habitat ranges from coastal areas to open ocean environments.

Aim

The aim of a practical exercise is focused on the identification of the order: Lamniformes.

Systematic Classification

Kingdom: Animalia

Phylum: Chordata

Class: Chondrichthyes

Order: Lamniformes

Family: Alopidae and Lamnidae

Identification Key Characters of Family – Alopidae (Theresher Sharks)

- **Body Structure:** Large size; cylindrical trunk and precaudal tail, not depressed, and lacking lateral ridges.
- **Head:** Not laterally expanded or depressed; eyes on the sides without nictitating lower eyelids.
- **Snout:** Moderately long, bluntly conical, not flattened, without lateral teeth or barbels; nostrils well-separated from the mouth.
- **Mouth:** Small but arched and elongated, extending well behind the eyes; labial furrows present on the lower jaw or absent.
- **Teeth:** Small, blade-like, compressed, with erect to oblique cusps; anterior teeth in upper jaw slightly larger than lateral teeth.
- **Gill Slits:** Five small to medium-sized gill slits, last 2 behind pectoral-fin origins; no gill rakers on internal gill slits; minute spiracles present.
- **Fins:** Two dorsal fins without spines; first dorsal fin moderately large, high, angular, and much shorter than the caudal fin; second dorsal fin low and minute; anal fin present, very small.
- **Caudal Fin:** Strongly asymmetrical, upper lobe enlarged, about half the total length with a subterminal notch, undulated dorsal margin; lower lobe short but strong.
- **Caudal Peduncle:** Not depressed, without keels; precaudal pits present.

Layout

Identification Key Characters of Family – Lamnidae: (Mackerel Shark)

- **Size and Shape:** Large-sized sharks with a fusiform body.
- **Gill Slits:** Head with 5 gill slits, all in front of pectoral-fin origins; no nictitating eyelids.
- **Teeth:** Long and few in number, awl- or blade-like, with a single cusp; gill arches lack rakers.

- **Dorsal Fins:** Two dorsal fins; the first dorsal fin much shorter at the base than the caudal fin, positioned far in advance of pelvic fins.
- **Second Dorsal and Anal Fins:** Much smaller than the first dorsal fin, with narrow, pivoting bases.
- **Caudal Fin:** Lunate, less than 1/3 of total length; caudal peduncle strongly depressed dorsoventrally, expanded laterally, with prominent keels on each side extending well out on the caudal fin.
- **Intestinal Valve:** Ring type.
- **Color:** Back in various shades (pale grey, greyish blue, purplish blue, brownish, blackish grey, or black); underside white to lighter grey.

Layout

INFERENCE:

Chapter-10

Identification of Order – Carcharhiniformes

Introduction

This description characterizes members of the Carcharhiniformes order, commonly known as ground sharks. They typically exhibit cylindrical to slightly compressed trunks and conical to depressed heads. Gill slits numbering five pairs are situated on the sides of the head, with some species having dorsolateral gill slits. These sharks may possess spiracles and lateral or dorsolateral eyes with true nictitating lower eyelids. Their moderately large to very large mouths extend behind the anterior ends of the eyes. Usually lacking enlarged molariform posterior teeth, they have two dorsal fins without spines and moderate-sized to large pectoral fins. Pelvic fins range from small to moderately large, with a continuous vent. The caudal fin typically features a long dorsal lobe and a varying ventral lobe. Intestinal valves are of the spiral or scroll type.

Aim

The aim of a practical is exercise focused on the identification of the order: Carcharhiniformes.

Systematic Classification

Kingdom: Animalia

Phylum: Chordata

Class: Chondrichthyes

Order: Carcharhiniformes

Family: Scyliorhinidae, Proscyllidae, Leptochariidae, Triakidae, Hemigaleidae, Carcharhinidae and Sphyrinidae.

Identification Key Characters of Family – Scyliorhinidae (Cat Sharks)

- **Head:** Without laterally expanded blades; elongated and fusiform, oval, or slitlike eyes, length over 1.5 times the height; rudimentary nictitating eyelids; moderately large spiracles.
- **Nasal Flaps:** Anterior nasal flaps variably shaped, not barbel-like (except for Poroderma genus).
- **Nostrils:** Internarial width about 0.6 to 1.3 times nostril width.
- **Labial Furrows:** Absent or very short to very long.
- **Teeth:** Small, with acute narrow cusps, often with lateral cusplets and basal ledges; not bladelike; similar in both jaws; posterior teeth comblike or not; tooth rows range from 40 to 101/34 to 111.
- **Precaudal Pits:** Absent.
- **Dorsal Fins:** First dorsal fin small, not keel-like, much shorter than the caudal fin; positioned over or behind pelvic bases, its origin slightly ahead of pelvic origins in some cases.
- **Pectoral Fins:** Radials confined to bases of fins.
- **Caudal Fin:** Ventral caudal lobe absent or very weak; no undulations or ripples in dorsal caudal margin.

Layout

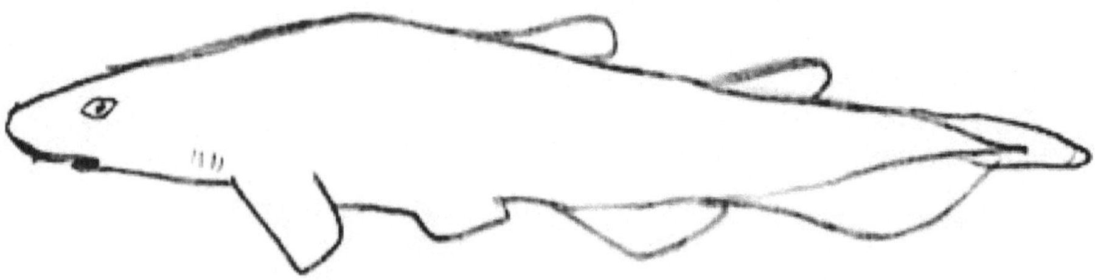

Identification Key Characters of Family – Proscyllidae (Finback catsharks)

- **Precaudal Pits:** Absent.
- **Dorsal Caudal Margin:** Not undulated.
- **Teeth:** Posterior teeth on dental bands are comblike.
- **Labial Furrows:** Very short or absent; if present, confined to mouth corners.

Layout

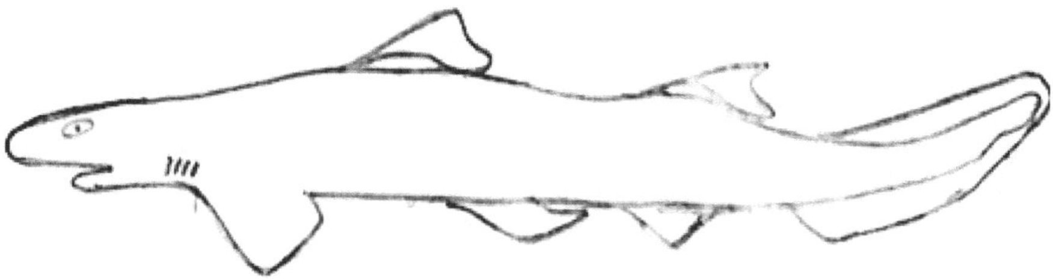

Identification Key Characters of Family – Leptochariidae (Barbeled Hound Sharks)

- **Anterior Nasal Flaps:** Formed as slender barbels.
- **Labial Furrows:** Upper labial furrows extremely long, nearly equal to internarial width and over half mouth width.
- **Intestinal Valve:** With 14 to 16 turns.
- **Neurocranium:** No supraorbital crests.

Layout

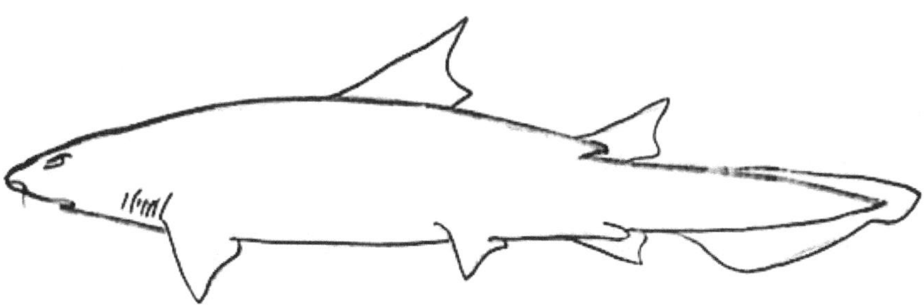

Identification Key Characters of Family – Triakidae (Hound Sharks)

- **Anterior Nasal Flaps:** Usually not barbel-like (except for Furgaleus).
- **Labial Furrows:** Upper labial furrows shorter, considerably less than internarial width and less than half of mouth width.

- **Intestinal Valve:** With 4 to 10 turns.
- **Neurocranium:** Supraorbital crests present.

Layout

Identification Key Characters of Family – Hemigaleidae (Weasel Shark)

- **Anterior Nasal Flaps:** Usually not barbel-like.
- **Labial Furrows:** Upper labial furrows shorter, considerably less than internarial width, and less than half of mouth width.
- **Intestinal Valve:** Typically with 4 to 10 turns. Intestinal valve of spiral type
- **Neurocranium:** Supraorbital crests present.

Layout

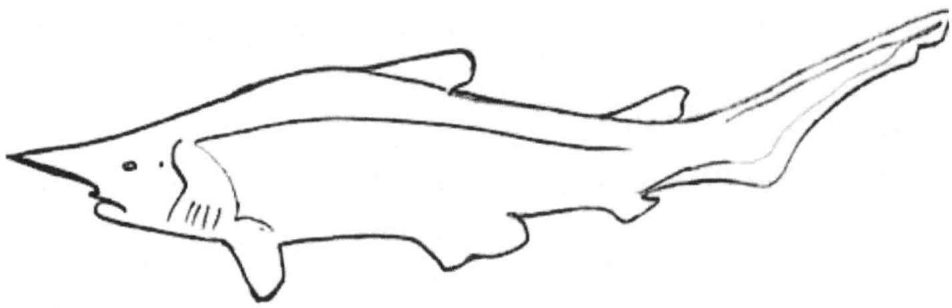

Identification Key Characters of Family – Carcharhinidae (Requiem sharks)

- **Body Shape:** Generally cylindrical or slightly compressed, with a fusiform shape.
- **Head:** Conical to depressed, usually not highly elongated.
- **Gill Slits:** Five pairs of gill slits on the sides of the head, with the last 1 to 3 over the pectoral fin bases.
- **Nostrils:** Usually lacking barbels and nasoral grooves, with circumnarial grooves absent.
- **Eyes:** Positioned laterally or dorsolaterally, with true nictitating lower eyelids.

- **Teeth:** Typically without enlarged molariform posterior teeth, and anterior teeth not separated by small intermediate teeth or a gap from lateral teeth.
- **Dorsal Fins:** Two dorsal fins, without spines.
- **Pectoral Fins:** Moderate-sized to large, not ray-like, without triangular anterior lobes.
- **Pelvic Fins:** Small to moderately large, with the vent continuous with inner margins.
- **Caudal Fin:** Characterized by a long dorsal lobe, with the ventral lobe varying from long to absent, and the vertebral axis raised into the dorsal caudal lobe.
- **Intestinal Valve:** scroll-type.

Layout

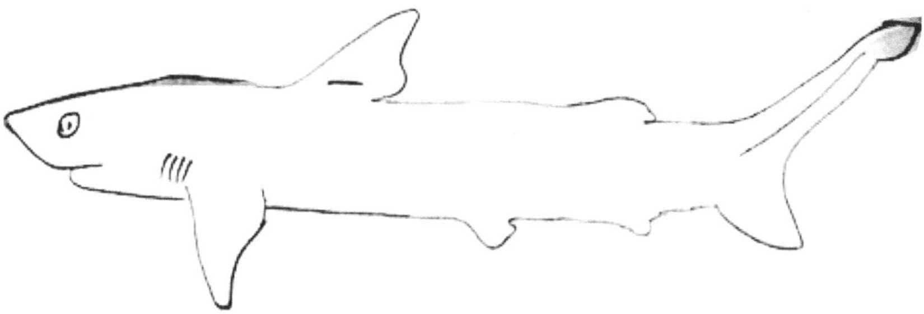

Identification Key Characters of Family : Sphyrnidae (Hammerheaded Sharks)

- **Body Shape:** Generally fusiform (spindle-shaped) or slightly compressed.
- **Gill Slits:** Five pairs of gill slits, similar to other shark families.
- **Nostrils:** Positioned at the ends of the cephalofoil extensions, without barbels or nasoral grooves.
- **Eyes:** Located laterally or dorsolaterally on the ends of the cephalofoil extensions.
- **Teeth:** Typically with sharp, triangular teeth, similar to other shark families.
- **Dorsal Fins:** Two dorsal fins, without spines.
- **Pectoral Fins:** Moderate-sized to large, not ray-like, without triangular anterior lobes.
- **Pelvic Fins:** Small to moderately large, with the vent continuous with inner margins.
- **Caudal Fin:** Characterized by a long dorsal lobe, with the ventral lobe varying from long to absent, and the vertebral axis raised into the dorsal caudal lobe.

Layout

INFERENCE:

Chapter-11

Identification of Order – Rajiformes

Introduction

The order Rajiformes comprises skates and rays, characterized by their flattened bodies and enlarged pectoral fins resembling wings. Found in various marine habitats worldwide, these creatures play vital roles as predators and prey, contributing to the balance of marine ecosystems. With diverse reproductive strategies and specialized feeding habits, they are integral parts of benthic communities.

Aim

The aim of a practical exercise focused on the identification of the order: Rajiformes.

Systematic Classification

Kingdom: Animalia

Phylum: Chordata

Class: Chondrichthyes

Order:s Rajiformes

Family: Rajidae and Rhinobatidae

Identification Key Characters of Family – Rajidae (Skates)

- Rounded or angular body.
- Laterally widened body, which can be interpreted as greatly flattened from side to side.
- Body and head greatly depressed, forming a rhomboidal disc.
- Pectoral fins fused with the body, contributing to the disc shape.
- Tail terminates in a blunt tip without a caudal fin.
- Presence of two dorsal fins located posteriorly (toward the rear) on the body.
- Dorsal fins are distinct and relatively small.
- They are positioned near the end of the caudal (tail) region.

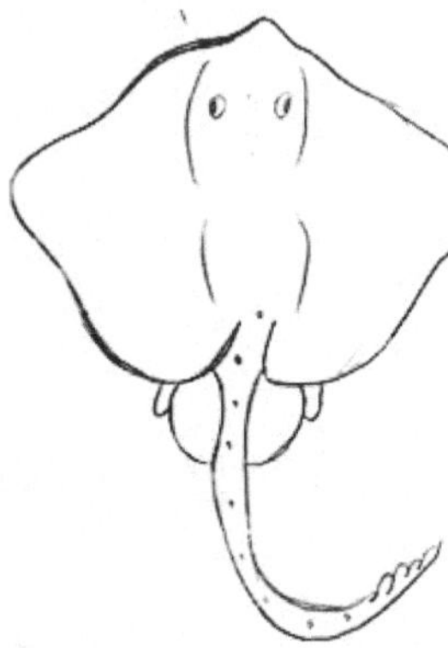

Layout

Identification Key Characters of Family – Rhinobatidae (Guitar fishes)

- Rather elongated body compared to other rays and skates.
- The snout is typically pointed and elongated.
- Broad, soft, and rounded snout, which is a distinctive feature of guitarfishes.
- The pointed snout is notably different from the flattened disc-shaped heads of typical rays.
- The first dorsal fin is triangular.
- Positioned anteriorly, usually before the base of the pelvic fins.
- Eyes are slightly smaller in proportion to the body and are entirely separated from spiracles.

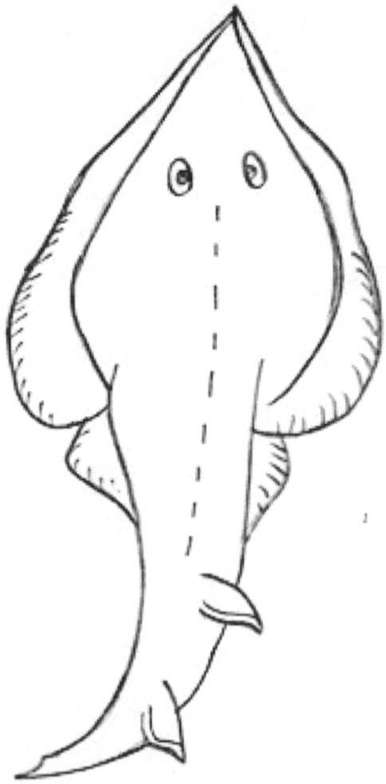

Layout

Chapter-12

Identification of Order – Rhinopristiformes

Introduction

Sawfish are a group of large cartilaginous fish known for their distinctive long, flat snouts, which are lined with sharp teeth resembling a saw. They belong to the order Rhinopristiformes, family Pristidae and are closely related to rays and sharks. Sawfish are typically found in shallow coastal waters and estuaries in tropical and subtropical regions around the world. They are formidable predators, using their saw-like rostrum to slash through schools of fish and crustaceans, stunning them before consuming their prey. Despite their impressive appearance, sawfish are considered critically endangered due to overfishing, habitat loss, and incidental capture in fishing gear. Conservation efforts are crucial to protecting these unique and ecologically important creatures and ensuring their survival in the wild.

Aim

The aim of a practical exercise focused on the identification of the order: Rhinopristiformes.

Systematic Classification

Kingdom: Animalia

Phylum: Chordata

Class: Chondrichthyes

Order: Rhinopristiformes

Family: Pristidae

Identification Key Characters of Family – Pristidae (Saw fish)

- Elongate body, not resembling the typical flattened disc shape of most rays.
- The body often appears slender and elongated compared to other batoids.
- The snout is elongated and produced.
- It is notable for its saw-like structure lined with teeth.
- The teeth along the rostrum are bony, prominent, and arranged in a saw-like pattern.
- The teeth at the posterior end of the rostrum extend well beyond the anterior two pairs of rostral teeth.
- These teeth are a defining feature of sawfish and are used for various functions, including hunting and defense.

Layout

INFERENCE:

Chapter-13

Identification of Order – Torpediniformes

Introduction

The Torpediniformes is an order of electric rays, also known as torpedoes or numbfishes, characterized by their ability to produce electric shocks for defense and hunting. These cartilaginous fishes are part of the subclass Elasmobranchii, which includes sharks and rays. Torpediniformes are typically found in coastal waters and deep-sea habitats worldwide, where they feed on a variety of prey including fish and invertebrates. Their electric organs, located on either side of their bodies, can produce electric discharges used to stun or deter predators and to capture prey.

Aim

The aim of a practical exercise is focused on the identification of the order: Torpediniformes.

Systematic Classification

Kingdom: Animalia

Phylum: Chordata

Class: Chondrichthyes

Order: Torpediniformes

Family: Torpedinidae and Narcinidae.

Identification Key Characters of Family – Torpedinidae (Electric rays)

- Typically, rounded or angular body.
- The body is laterally widened, often forming a disc-like shape.
- Notably, electric rays possess two large electric organs located in the front part of the disc, one on each side of the head.
- These organs generate electric charges that the ray uses for hunting, defense, and communication.
- The eyes of electric rays are relatively small compared to the size of their disc-shaped body.
- The dorsal surface often displays spots against a brown background, while the ventral (lower) side tends to be pale in coloration.

Layout

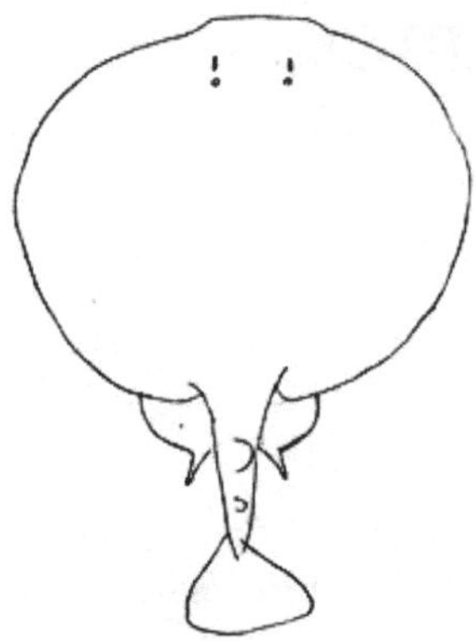

Identification Key Characters of Family – Narcinidae (Numb fishes)

- **Size and Body Shape:** Torpediniformes are typically small to medium-sized batoids, ranging from 15 to 66 cm in total length, with a stout shark-like tail that is equal to or longer than the disc length. They have large oval or rounded pectoral discs.
- **Skin and Body Structure:** Their bodies lack dermal denticles or thorns, and they have a depressed and flattened trunk, distinct from shark-like body structures. The precaudal tail is moderately depressed and abruptly narrower than the trunk, without a barbed sting on the dorsal surface.
- **Head and Snout:** Torpediniformes have broad and depressed heads with moderately elongated, broadly rounded or obtusely rounded-angular snouts. They possess a large rostral cartilage supporting the snout but lack saw-like structures or lateral saw teeth.
- **Gills and Mouth:** They have five small gill openings on the underside of the front half of pectoral-fin bases. The mouth is transverse and straight,

with strong labial folds and nostrils located just anterior to it, connected by broad nasoral grooves.

- **Teeth:** Their oral teeth are small, with rounded or oval bases and short to moderately long cusps. Tooth bands remain exposed when the mouth is closed, and there are typically between 12 and 33 exposed rows in either jaw (in adults).

Layout

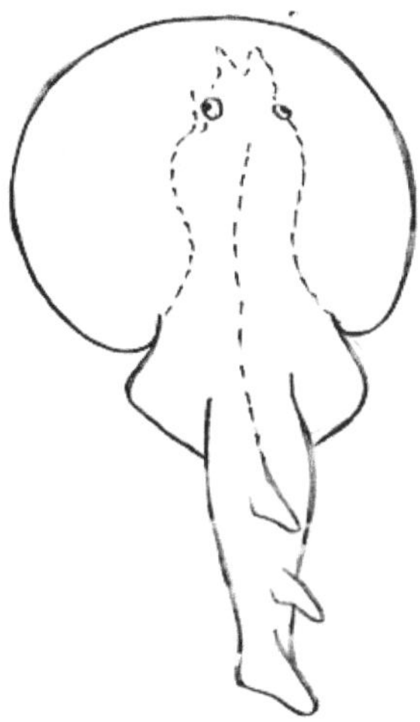

INFERENCE:

Chapter-14

Identification of Order – Myliobatiformes

Introduction

The Myliobatiformes (Eagle rays, Devil Rays, Stingrays) are the second largest group of batoids and are characterized by their enlarges, expanded pectoral fin.

Aim

The aim of a practical exercise focused on the identification of the order: Myliobatiformes.

Systematic Classification

Kingdom: Animalia

Phylum: Chordata

Class: Chondrichthyes

Order: Myliobatiformes

Family: Gymnuridae, Dasyatidae, Myliobatidae and Mobulidae.

Identification Key Characters of Family – Gymnuridae (Butterfly rays)

- Medium to large-sized stingrays, with a disc width exceeding 2 meters.
- The body is strongly depressed, forming a rhomboid disc shape with the head, trunk, and expanded pectoral fins.

- ❑ The disc is notably broader than it is long, at least 1.5 times wider than long.

Layout

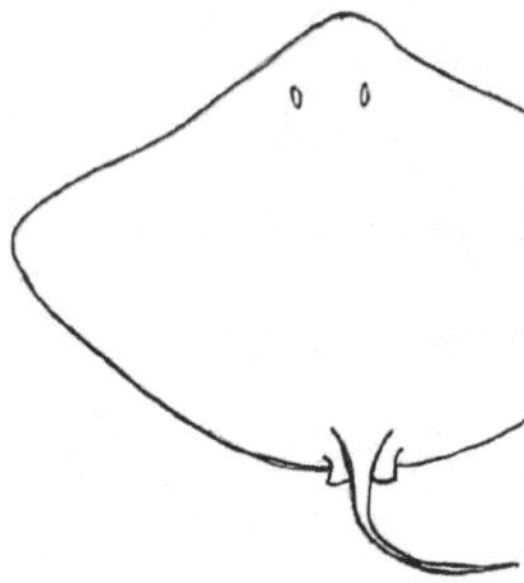

Identification Key Characters of Family – Dasyatidae (Big skates)

- ❑ Rounded or angular body laterally widened.
- ❑ The disc is at most 1.3 times as broad as long.
- ❑ Tail significantly longer than the disc width.
- ❑ Mouth floor with several fleshy papillae (protruding, nipple-like structures).
- ❑ Caudal fin thin, and the dorsal fin may be feeble or absent.
- ❑ One or two serrated spines present in the tail.
- ❑ Normal snout not notably elongated or modified.
- ❑ Head is not elevated, maintaining a relatively flat position.

Layout

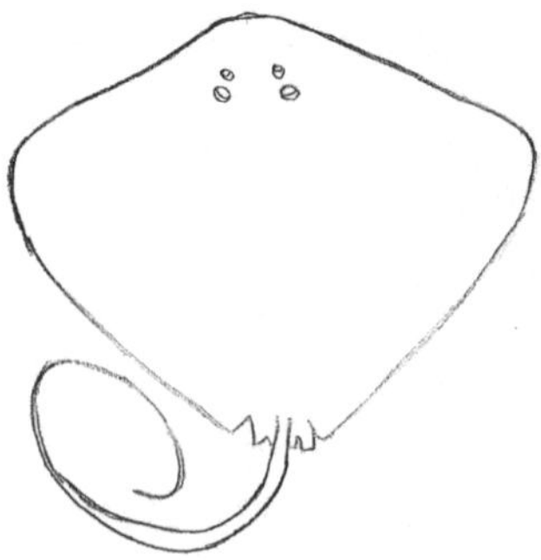

Identification Key Characters of Family – Myliobatidae (Eagle rays and Cownose rays)

- ❑ Rounded or angular body, laterally widened.
- ❑ Typically, a more pronounced disc shape with a wider span.

- ❑ Caudal fin thin, and the dorsal fin is often feeble or entirely absent.
- ❑ Their tails are whip-like and notably longer than the length of their bodies.
- ❑ The head is elevated above the pectoral fins, giving a somewhat raised appearance to the front of the ray's body.

Layout

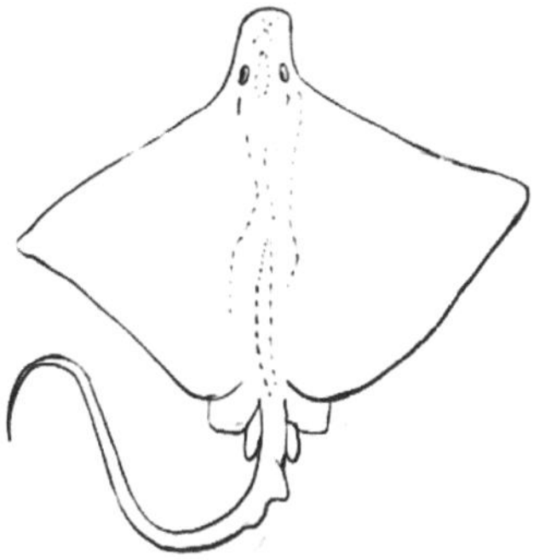

Identification Key Characters of Family – Mobulidae (Devil rays and Manta rays)

- ❑ Rounded or angular body laterally widened.
- ❑ The body form is typically broader.
- ❑ The pectoral fins are highly enlarged and wing-like, contributing to their distinct appearance.
- ❑ Like other rays, they have a thin caudal fin, and the dorsal fin is typically feeble or absent.
- ❑ The snout is produced as a fleshy flap on each side, which is a distinguishing feature.
- ❑ These rays have distinct modified cephalic (head) organs that are often present in pairs and contribute to their unique sensory capabilities.

Layout

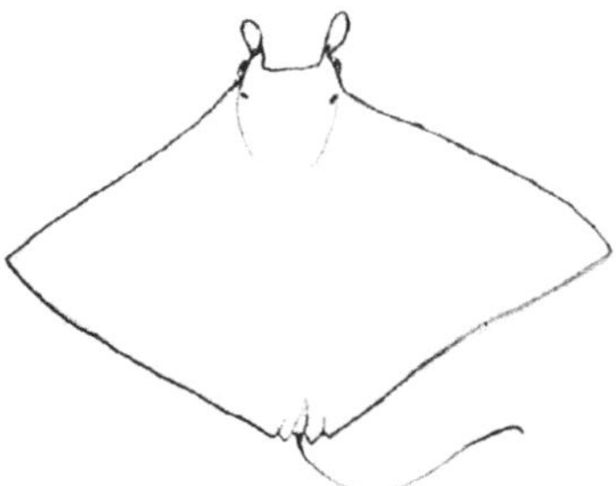

INFERENCE:

Chapter-15

Identification of Order – Clupeiformes

Introduction

Clupeiformes is an order of ray-finned fishes that includes some of the most important fish species in commercial fisheries and ecosystems worldwide. Commonly known as herrings, sardines, and anchovies, these fishes are characterized by their streamlined bodies, forked tails, and silver-colored scales. They are found in both marine and freshwater habitats, ranging from coastal waters to open oceans and inland lakes and rivers.

Aim

The aim of a practical exercise is focused on the identification of the order: Clupeiformes.

Systematic Classification

Kingdom: Animalia

Phylum: Chordata

Class: Actinopterygii

Order: Clupeiformes

Family: Clupeidae, Engraulidae and Chirocentridae.

Identification Key Characters of Family – Clupeidae (Herrings, Sardines, Shads and Menhadans)

- ❑ Body shape varies from fusiform (spindle-shaped) to either round or strongly compressed.
- ❑ Lack of scales on the head.
- ❑ Jaw teeth, if present, are typically small or minute.
- ❑ Belly scutes present.
- ❑ Single dorsal fin, usually small and positioned near the midpoint of the body.
- ❑ Pelvic fins are generally located more or less below the base of the dorsal fin.
- ❑ In some species, dorsal and pelvic fins may be absent.
- ❑ Fins consist of soft rays only.
- ❑ Lateral line may extend a few scales behind the head in some species, but it might be absent in others.
- ❑ Scales are typically cycloid, meaning they are smooth to the touch.
- ❑ Usually, there are 5-10 branchiostegal rays.

Layout

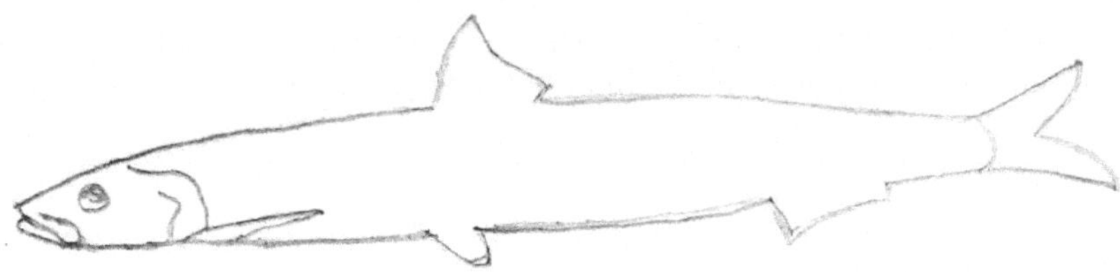

Identification Key Characters of Family – Engraulidae (Anchovies)

- ❑ **Size:** Typically, small to moderate-sized (10 to 20 cm standard length, occasionally larger).
- ❑ **Snout:** Prominent pig-like projection beyond the lower jaw tip.
- ❑ **Jaws:** Slender, long lower jaw extending past the eye.
- ❑ **Teeth:** Usually small; in Lycothrissa, they're canine-like.
- ❑ **Eyes:** Large, covered by an adipose eyelid.
- ❑ **Gill Rakers:** Short and not numerous in Indo-Pacific genera.
- ❑ **Pelvic Scute**: Always has ascending lateral arms.

- **Scutes**: Indo-Pacific species often have pre- and post pelvic scutes and a spine-like scute anterior to the dorsal-fin origin.
- **Fins:** Single, short dorsal fin near body midpoint. No adipose fin. Pectoral fins set low. Pelvic fins with specific soft ray counts.
- **Anal Fin**: Moderate length, varying in soft fin ray count.
- **Scales:** Cycloid, moderate in number (about 30 to 60 in lateral series), often shed when captured.
- **Coloration:** Blue-green dorsum, silvery flanks, occasional silver stripe or dark saddle on the nape. Fins and mouth can vary in color, with occasional black markings but no flank spots as seen in clupeids.

Layout

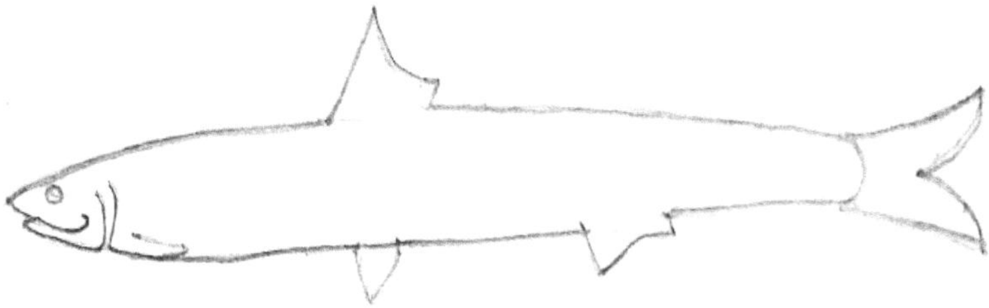

Identification Key Characters of Family – Chirocentridae (Wolf herring)

- Body elongate and compressed. Jaws with fanglike teeth (Canine).
- Intestine with spiral valve.
- Gill rakers 17-22.
- Pyloric caeca absent.
- Scales small.
- Dorsal fin rays 16-18.
- Pectoral fin rays 13-15. Branchiostegal rays 6-8. Cauda fin deeply forked.

Layout

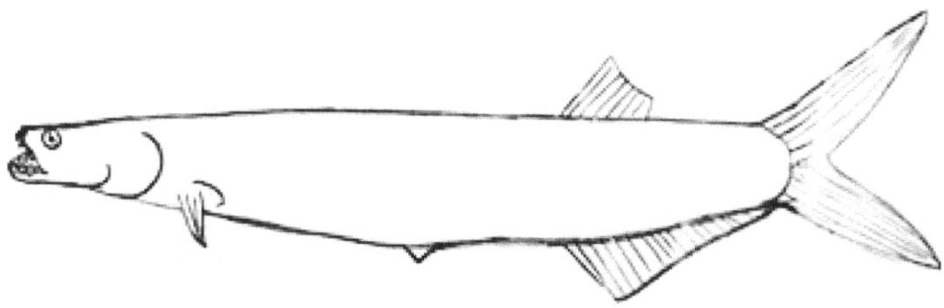

INFERENCE

Chapter-16

Identification of Order – Perciformes

Introduction

Perciformes, one of the largest and most diverse orders of ray-finned fishes, encompasses a vast array of species found in both marine and freshwater habitats across the globe. With a habitat preference mostly inclined towards marine environments, Perciformes exhibit remarkable adaptability, occupying various niches from shallow coastal waters to the depths of the open ocean.

Aim

The aim of a practical exercise focused on the identification of the order: Perciformes.

Suborder: Percoidei

- Maxillary bones not included in mouth gape; tooth-bearing pre-maxilla
- Two or single dorsal fins with sharp spines followed by rays
- Pelvic fin with single spine and rays placed at posterior part of body profile

Systematic classification

Kingdom: Animalia

Phylum: Chordata

Class: Actinopterygii

Order: Perciformes

Suborder: Percoidei

Family: Serranidae, Siliaginidae, Lactaridae, Rachycentridae, Echeneidae, Coryphaonidae, Gerridae, Mullidae , Haemulidae, Sparidae, Lethrinidae, Nemipteridae, Polynemidae, Sciaenidae, Carangidae, Monodactylidae, Chaetodontidae, Drepanidae, Pomacanthidae and Terapontidae, Leiognathidae, caesionidae

Identification Key Characters of Family – Serranidae (Groupers)

- **Operculum Structure:** Bearing Three Spines (Main Spine with One Above and One Below)
- **Lateral Line:** Complete and Continuous, Excluding One Species (Lacking in One Species)
- **Dorsal Fin:** Notched in Some Cases, Comprising 7-12 Spines
- **Anal Fin:** Consisting of Three Spines
- **Caudal Fin:** Typically, Rounded, Truncate, or Lunate; Rarely Forked
- **Maxilla Position:** Tip Exposed Even with Closed Mouth
- **Pelvic Fin:** One Spine with 5 Soft Rays, No Scaly Axillary Pelvic Process
- **Branchiostegal Rays:** Usually 7 in Number
- **Vertebrae Count:** Ranging Between 24-26
- **Reproductive Characteristics:** Monoecious with Some Functional Hermaphrodites; Protogynous Hermaphrodites in Groupers

Layout

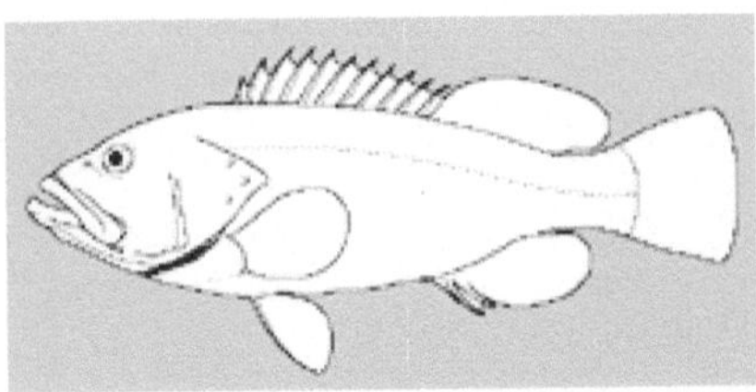

Identification Key Characters of Family – Sillaginidae (Smelt and Whitning)

- **Body Shape:** Sillaginidae species have elongated, slender bodies with a cylindrical or slightly compressed shape.
- **Mouth and Barbels:** They typically possess terminal mouths with small teeth or tooth plates and various lengths of barbels around their mouths to aid in locating food.
- **Fins:** Sillaginidae have a single dorsal fin closer to the head than the tail, and well-developed pectoral fins that assist in maneuverability and stability.
- **Coloration and Pattern:** Their coloration ranges from plain to mottled, often in shades of brown, gray, or silver, with some species featuring darker spots or markings for camouflage.

Layout

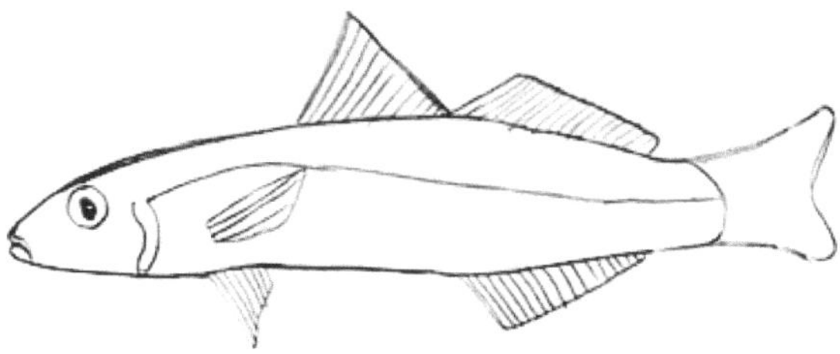

Identification Key Characters of Family – Lactariidae (False travelly)

- Dorsal fins 2.
- Soft-rayed parts of dorsal and anal fins covered with deciduous scales (all scales easily shed).
- Anal fin base is longer than that of second dorsal fin
- Caudal fin forked
- Upper and lower jaws each with 2 small front canine teeth at front. 24 vertebrae.
- Mouth large and oblique, with a prominent lower Jaw.

Layout

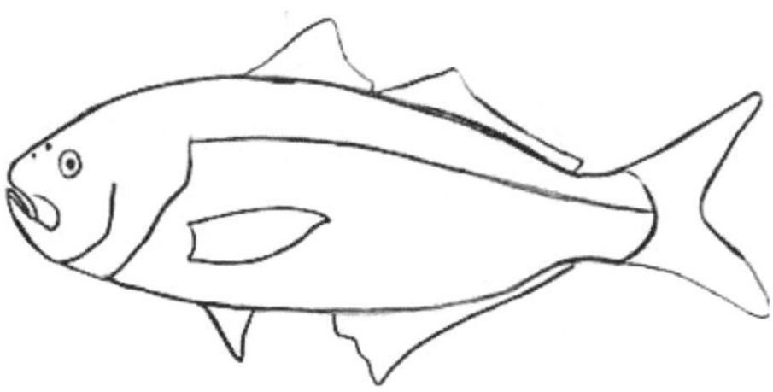

Identification Key Characters of Family – Echeneidae (Remoras and Suckers)

- ❑ **Body Structure:** Elongate with Flattened Head
- ❑ **Sucking Disc Description:** Transverse Movable Lamina
- ❑ Evolution of Suckssing Disc from Dorsal Fin
- ❑ Jutted Mandible
- ❑ Small and Cycloid Scales
- ❑ Absence of Spines in Dorsal and Anal Fins
- ❑ Soft Ray Count in Dorsal and Anal Fins
- ❑ Lack of Swim Bladder

Layout

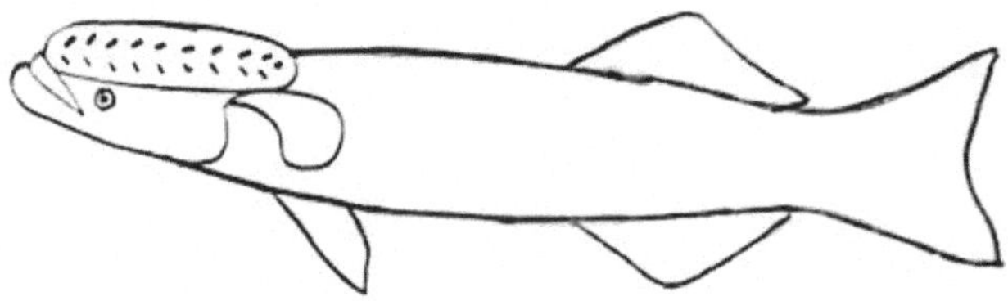

Identification Key Characters of Family – Rachycentridae (Cobias)

- ❑ Projected Lower Jaw
- ❑ **Body Shape:** Elongate with Depressed Head
- ❑ **Distinctive Body Markings:** Three Dark Lateral Stripes
- ❑ First Dorsal Fin Description- Short, Strong, Isolated Spines
- ❑ Arrangement of Spines preceding Dorsal Fin
- ❑ **Dorsal Fin Structure:** Spines and Soft Rays Count
- ❑ **Characteristics of the Anal Fin:** Spines and Soft Rays Count
- ❑ Absence of Membrane Connection
- ❑ Lunate Caudal Fin Shape in Adults
- ❑ Upper Lobe Longer than Lower
- ❑ Wavy Pattern in the Front Part

Layout

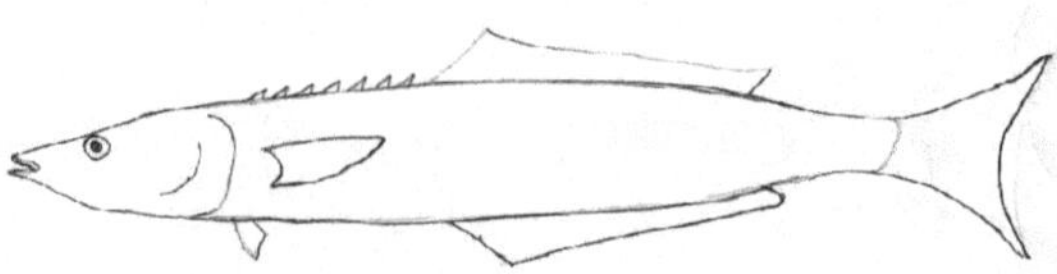

Identification Key Characters of Family – Coryphaenidae (Dolphin fishes)

- ❑ Slender, Compressed Head and Body Structure
- ❑ Description of the Single Dorsal Fin
- ❑ Extends from Head to Nearly Full Length of Body
- ❑ Soft Ray Count: 48-65
- ❑ Lack of Spines in Dorsal and Anal Fins
- ❑ Deeply Forked Caudal Fin
- ❑ Beautiful Coloration in Live Specimens
- ❑ Curved Lateral Line Above Pectoral Fin
- ❑ Extraordinary Length of Dorsal and Anal Fins, Almost Reaching Caudal Fin

Layout

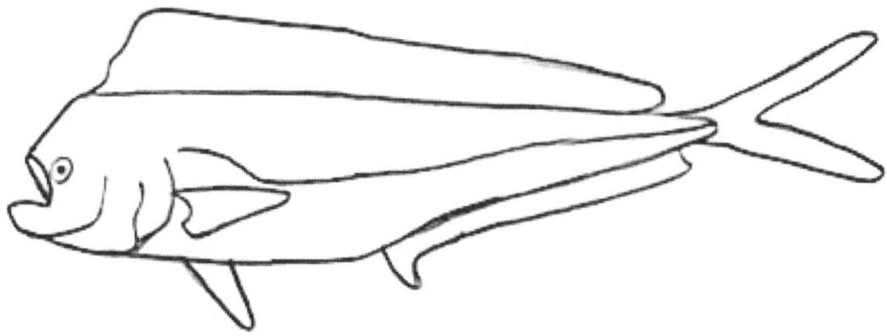

Identification Key Characters of Family – Leiognathidae (Slimys, slipmouths, or ponyfishes)

- ❑ **Body Structure:** Strongly Compressed and Slimy
- ❑ Characteristics of Scales and Head
- ❑ Small Scales
- ❑ Naked Head with Bony Ridges on Upper Surface
- ❑ United Gill Membranes with Isthmus
- ❑ Small, Very Protractile Mouth
- ❑ Lack of Palatal Teeth
- ❑ Absence of Pseudobranchiae
- ❑ **Dorsal and Anal Fin Structure:** Spines and Soft Rays Count
- ❑ Locking Mechanism in Dorsal and Anal Fin Spines
- ❑ Presence of Scaly Sheath at Fin Bases
- ❑ Vertebrae Count: 22-23
- ❑ Presence of Esophageal Luminous Organs
- ❑ Mucus Production (Reflecting the Common Name)

Layout

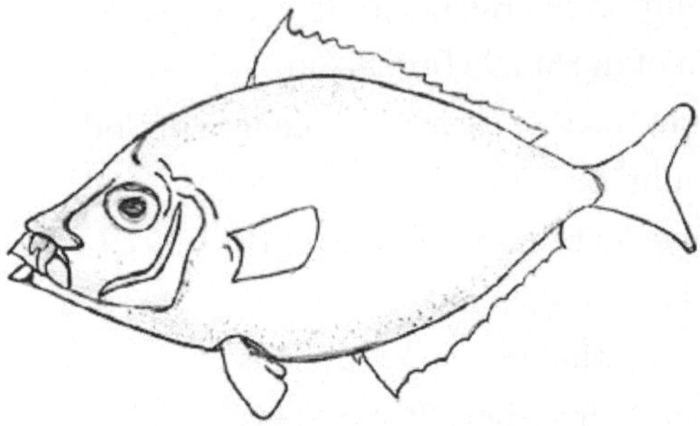

Identification Key Characters of Family – Lutjanidae (Snappers)

- Description of Dorsal Fin- Continuous or Slightly Notched, Spine Count: 10-12, Soft Rays: 10-17
- Anal Fin Composition-Three Spines, Soft Rays: 7-11
- Pelvic Fins Originating Behind Pectoral Base
- Moderate to Large, Terminal Mouth
- Enlarged Canine Teeth in Jaws
- Small Palatine Teeth, Vomer Description
- Maxilla Covered by Preorbital with Closed Mouth
- Presence and Absence of Scales on Different Facial Areas
- Absence on Anterior Head (Snout and Preorbital)
- Presence on Cheek, Preopercle, and Gill Cover in Rows

Layout

Identification Key Characters of Family – Caesionidae (Fusiliers)

- **Body Shape:** Oblong to Fusiform
- **Dorsal Fin:** Spine and Soft Ray Counts
- **Caudal Fin Structure:** Distinctly Forked with Pointed Lobes

- **Scale Distribution:** Horizontal Rows on Body, Exceptions in Gymnocaesio gymnoptera and Dipterygonotus balteatus
- Mouth and Dentition:
- Small Highly Protrusible Mouth
- **Description of Teeth:** Small or Minute Conical
- Coloration and Markings:
- Axil of Pectoral Fins-Black Marking
- Variations inside Coloration and Caudal Fin Markings

Layout

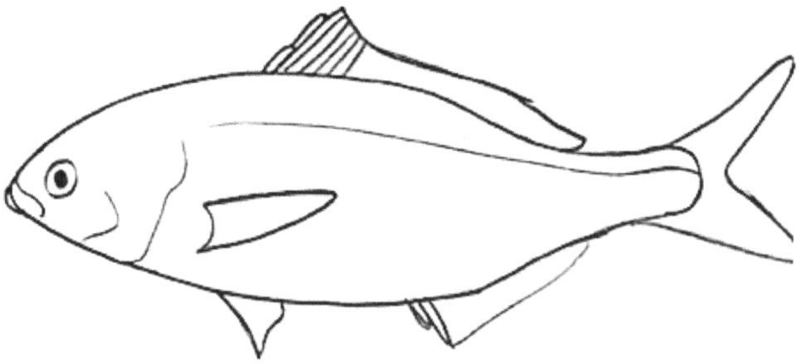

Identification Key Characters of Family – Gerreidae (Silver biddies and Mojarras)

- Highly Protractile Mouth
- Scaly Head Structure with Smooth Upper Surface
- Presence of Scaled Sheath Along the Base of Dorsal and Anal Fins
- Gill Membranes Not United to Isthmus
- Description of Tail: Deeply Forked
- Vertebrae Count: 24
- Maximum Attainable Length: Up to 35 cm in Gerres filamentosus.

Layout

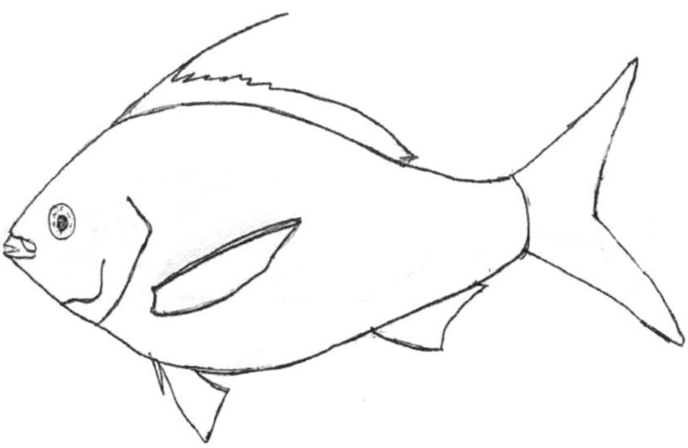

Identification Key Characters of Family – Mullidae (Goat fishes)

- **Body Structure:** Elongated
- **Arrangement of Dorsal Fins:** Far Apart
- **First Dorsal Fin:** 6-8 Spines
- **Second Dorsal Fin:** One Spine, 8-9 Soft Rays (Shorter than Anal Fin)
- **Anal Fin Composition:** 1-2 Spines, 5-8 Soft Rays
- **Caudal Fin Shape:** Forked
- **Vertebrae Count:** 24
- **Chin Description:** Two Long Barbels
- **Functions:** Contain Chemosensory Organs, Used for Probing Sand or Reef Holes for Prey
- Bright Coloration in Many Species

Layout

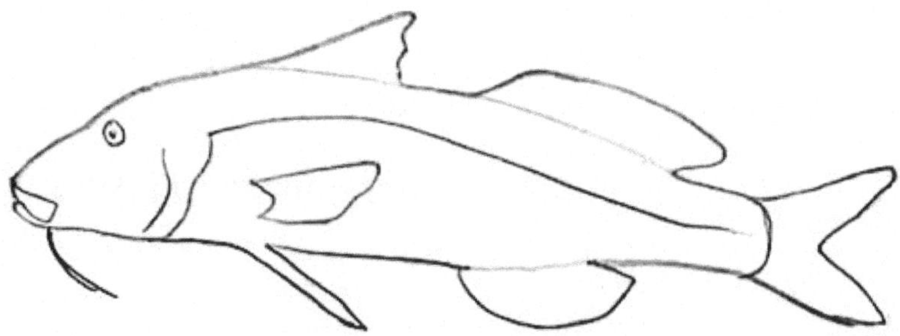

Identification Key Characters of Family – Haemulidae (Grunts, sweetlips, Rubber lips and Hotlips)

- Dorsal fin continuous, 9-14 spines, 11-26 soft rays.
- Anal fin 3 spines, 6-18 soft rays. Small mouth with thick lips.
- Usually cardiform jaw teeth, with vomer generally toothless.
- Usually with enlarged chin pores.
- Branchiostegal rays 7. Vertebrae 26 or 27 (10 or 11 + 16).
- Scales present on entire head (Except front of snout, tips and chin.
- Chin with 2 pores anteriorly and a median pit or 6 pores and no pit in Ptectorhincus
- Preopercle with a slightly concave and serrated posterior margin.
- Lips thick, tip of upper jaw hidden when mouth closed.

Layout

Identification Key Characters of Family – Sparidae (Porgies, Seabreams, Stumpnoses)

- Head large, often with a steep upper profile, cheeks scaly, preopercle with or without scales, without spines or serrations on margin
- Dorsal fin single, with 10 to 13 spines and 9 to 17 soft rays; no notch between spiny and soft portions
- Anal fin with 3 spines and 7 to 15 soft rays
- An axillary scale at the base of pelvic fins

Layout

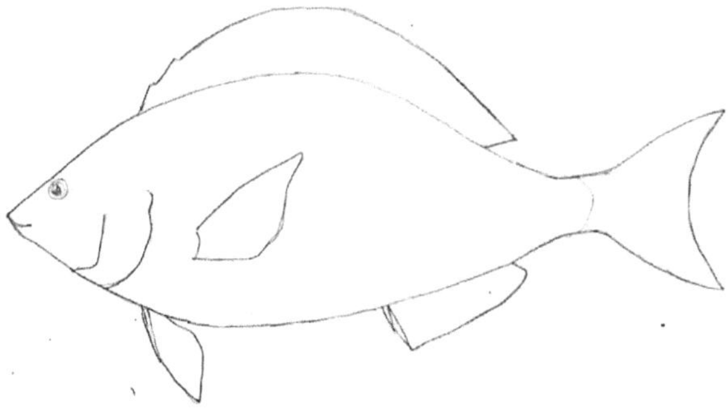

Identification Key Characters of Family – Lethrinidae (Emperors, Pig face bream, Large eye breams)

- Short gill rakers that are knob-like.
- The dorsal fin is single and continuous, with 10 spines and 9 or 10 branched (soft) rays.
- Cheeks, upper surface of the head, and preorbital area are scaleless in the genus Lethrinus, while other genera in Lethrinidae have scales on the cheek.

- ❑ The maxilla overlaps the premaxilla.

Layout

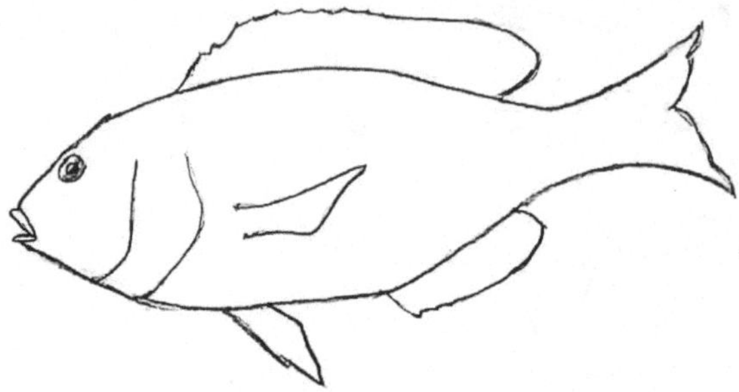

Identification Key Characters of Family – Nemipteridae (Thread fin breams / Monocle breams / Dwarf monocle breams)

- ❑ Single dorsal fin originating above pectoral fin
- ❑ bases, the first spine sometimes prolonged into a filament
- ❑ Just above pelvic fin, medium sized axillary scale present
- ❑ Caudal fin forked or emarginated, upper lobe sometimes with a filament Dorsal fin with 10 spines; soft rays 9.
- ❑ Three spines in anal fin; soft rays 7 (Nemipterus virgatus with 8).
- ❑ A filament off upper lobe of caudal fin in some species.
- ❑ Well developed subocular shelf and accessory pelvic keel.
- ❑ Intercalar present.
- ❑ A backward pointing spine below eye in Scolopsis and in some Parascolopsis species
- ❑ Front of head scaleless, scales on top of head (predorsal scales) beginning above or just in front of Eye.
- ❑ Cheeks with 3 rows of scales in Nemipterus, 4 to 7 rows in Scolopsis and Parascolopsis.

Layout

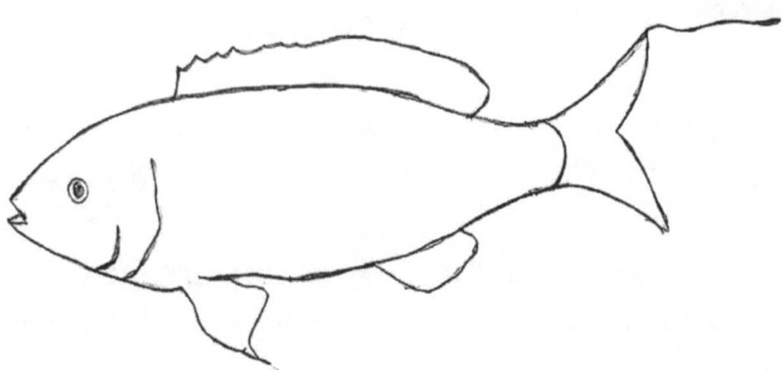

Identification Key Characters of Family – Polynemidae (Thread fins/ Tassel fishes)

- Mouth inferior. Pectoral fin with 2 sections.
- Upper section of pectoral fin with the rays attached; the lower section having 3-7 (except in Polistonemus with 14 or 15) long free rays.
- Spinous and soft dorsal fins far apart. Pelvics subabdominal.
- One spine in pelvic fin; soft rays 5. Deeply forked caudal fin.
- Lateral line extending upto caudal fin.

Layout

Identification Key Characters of Family – Sciaenidae (Croakers, Drums, Meagres)

- Except at the extreme part of the snout, both head and body are usually scaled.
- Present on various parts of the snout - rostral pores on the tip of the snout, marginal pores on the lower edge of the snout, and mental pores on the chin.
- Typically, there are 3 to 5 rostral pores on the top of the snout, 5 nearer the anterior margin of the mouth, and 3 pairs on the lower jaw.
- Bottom feeders such as some species of Johnius have well-developed rostral and mental pores.
- Mid-water feeders like Otolithes and Argyrosomus tend to have indistinct pores.
- Terminal mouth in species like Otolithes and Pennalia.
- Subterminal in Protonibea.
- Inferior or lower jaw projecting in Johnius and Umbrina.
- Dorsal fin is long and continuous, often with a deep notch between the anterior (spinous) and posterior (soft) portions.
- Anal fin typically has 2 spines, with the second spine enlarged.
- The lateral line extends to the hind margin of the caudal fin.

- Scales may vary, with cycloid scales on the head/ctenoid scales on the body, or only cycloid scales in certain species like *Jhonius dussumieri.*
- Internal features such as the shape and size of otoliths (ear bones) and swim bladder can be helpful in identifying genera/species.

Layout

Identification of Family : Carangidae (Jacks / Trevallies / Scads / Queenfishes / Amberjacks / Pilotfishes / Pampanos / Leerfish)

- In small juveniles, these fish have two separate dorsal fins.
- The first dorsal fin is moderate in height or very low.
- The anterior lobe of the second dorsal fin can range from scarcely produced to extremely long.
- The anal fin typically has two anterior spines (detached spines).
- The anterior lobe of the anal fin is described as low to elongate in shape.
- Scutes, which are enlarged, thickened, and often pointed scales in the lateral line, are a prominent feature in these fish.
- However, in some species, these scutes may be reduced.
- In certain genera, they might be entirely absent.

Layout

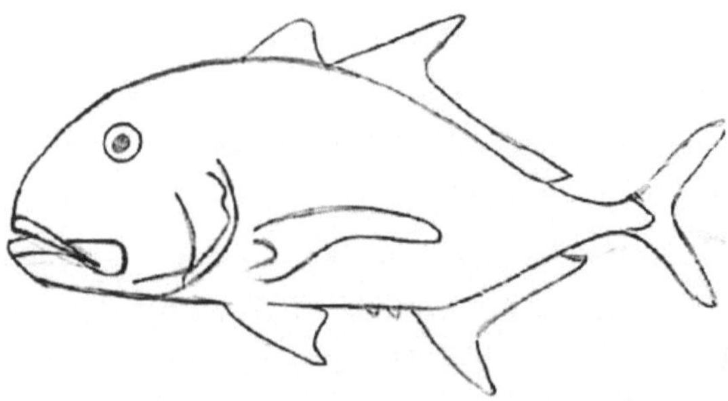

Identification of Family: Monodactylidae (Moonyfishes or fingerfishes)

- Body deep and highly compressed.
- Pelvic fins present in juveniles, lacking or vestigial in adults in Monodactylus.
- Dorsal fin with the base long and scaly; 5-8 short and graduated spines.
- Anal fin base long; 3 spines. Scales cycloid or ctenoid. Often silvery.

Layout

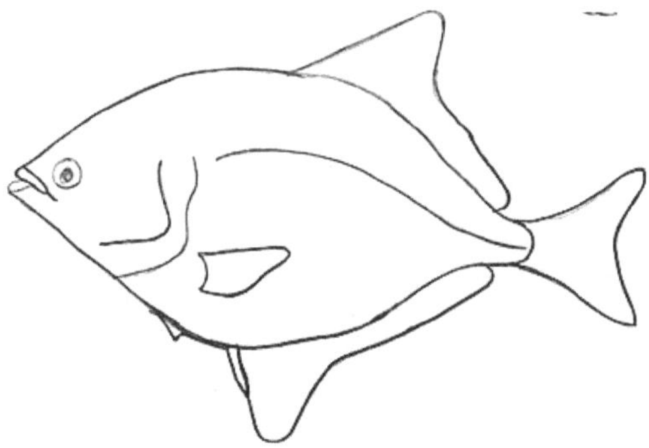

Identification of Family : Chaetodontidae (Butterfly fishes)

- Highly compressed body shape.
- Typically deep and strongly compressed body structure.
- Snout can range from slightly produced to greatly elongated.
- Dorsal fin is continuous, consisting of 6-16 spines and 15-30 soft rays.
- Anal fin has 3-5 (usually 3) spines and 14-23 soft rays.
- Caudal fin possesses 15 branched rays and is rounded to emarginate in shape.
- Scales extend onto both the anal and dorsal fins.
- Small, terminal, and protrusible mouth.
- Often characterized by a band or rows of small brushlike teeth.
- Gut typically coiled several times.
- Swim bladder with two anteriorly directed processes.
- A total of 24 vertebrae, organized as 11+13.
- Most species exhibit bright coloration.
- A dark band across the eye is a common feature.
- Presence of an 'eyespot' dorsally is typical.
- Axillary scaly process present at the base of the pelvic fin spine.
- Dorsal fin is continuous, sometimes with a slight notch between the spinous and soft portions.
- Possession of ctenoid scales.

Layout

Identification of Family: Drepanidae (Sickle fishes)

- Body deep and laterally compressed. Mouth markedly protractile.
- Spinal portion of dorsal fin with 13-14 spines, distinct from soft-rayed portion with 19-22 soft rays; anal fin with three spines and 17-19 soft rays; pectoral fins longer than head, falcate; 24 vertebrae.
- Maxilla distally exposed. Subocular shelf absent.

Layout

Identification of Family: Pomacanthidae (Angelfishes)

- Elongate body - oval to orbicular in shape/ strongly compressed
- Mouth protrusible.
- Dorsal fin single, continuous without a notch
- A strong spine at an angle in preopercle.

Layout

Identification of Family: Terapontidae (Terapon-perches / Terapons).

- ❑ Preopercle serrate/serrations more prominent in juveniles
- ❑ Two opercular spines; lower one larger and stronger; post temporal bone exposed in some species, its posterior margin serrate
- ❑ Single dorsal fin; with 11 to 14 spines
- ❑ Anal fin with 3 strong spines.

Layout

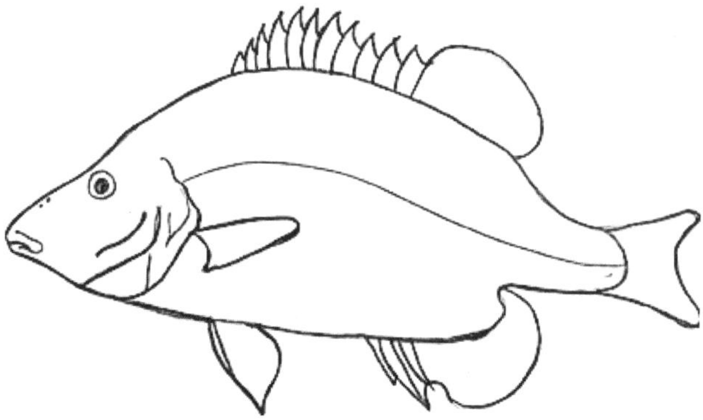

Sub order : Labroidei

- ❑ Well-developed teeth
- ❑ Parrot-like appearance in some species
- ❑ Large and smooth scales
- ❑ Compressed body, often colourful

Systematic classification

Kingdom: Animalia

Phylum: Chordata

Class: Actinopterygii

Order: Perciformes

Suborder: Percodei

Family: Pomacentridae, Labridae and Scaridae.

Identification of Family: Pomacentridae (Damsel fishes)

- **Nostrils:** Typically, these fish have one nostril on each side of their head. However, some species within the genera Chromis and Dascyllus might possess double nostrils.
- **Body Shape:** Damselfishes usually have a deep and compressed body structure, which aids in their maneuverability within reefs and rocky areas.
- **Mouth Size:** They typically have small mouths, adapted for feeding on small invertebrates and algae.
- **Lateral Line:** Their lateral line is often incomplete and interrupted.
- **Anal Fin:** The anal fin usually contains 2 spines, rarely 3.
- **Teeth:** Damselfishes generally lack palatine teeth.
- **Size:** Maximum length for these fish species is around 35 centimeters.
- **Coloration:** The coloration of damselfishes can vary significantly among individuals and can also be influenced by the environment or locality where they reside.

Layout

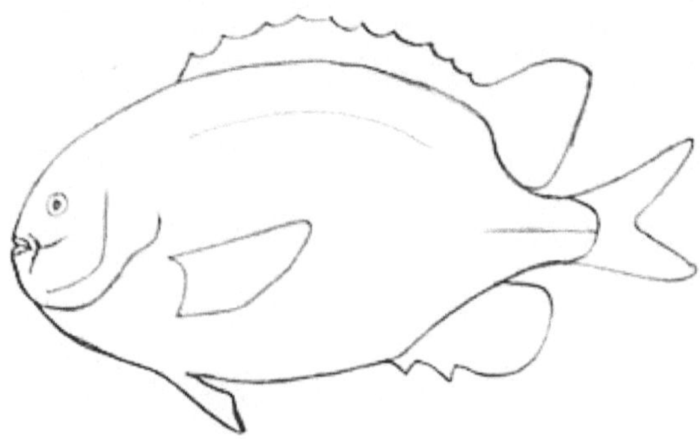

Identification of Family: Labridae (Wrasses, Hogfishes, Razor fishes, Coris, Tuskfishes)

- **Dorsal and Anal Fin Structure:** The dorsal fin typically consists of 8-21 spines (often fewer than 15) and 6-21 soft rays. The anal fin generally has 4-6 spines (sometimes 3) and 7-18 soft rays.
- **Scales and Vertebrae:** They possess cycloid scales, which are generally moderate to large in size, ranging from 25 to 80 along the side. The number of vertebrae usually falls between 23 and 42.
- **Lateral Line:** The lateral line can be interrupted or continuous. It may have a smooth curve or an abrupt curve below the soft portion of the dorsal fin, either being continuous or interrupted below the posterior part of the dorsal fin.
- **Snout Variation:** The genus Gomphosus within this family tends to have an elongated snout.
- **Size, Shape, and Color:** Wrasses exhibit extensive variation in size, body shape, and coloration. This diversity is often noticeable within the same species across different stages of their life cycle or based on gender differences.
- **Mouth Structure:** Their mouths are terminal, with prominent lips, suited for their varied feeding habits.

Layout

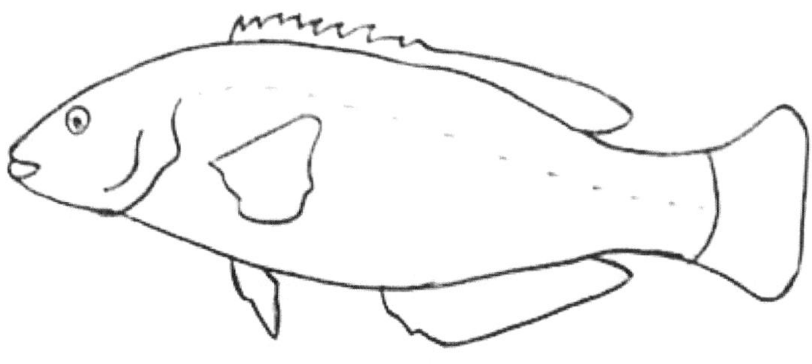

Identification of Family: Scaridae (Parrot fishes)

- **Head Shape:** Parrotfishes also have a bluntly rounded anterior head structure, similar to triggerfish.
- **Teeth Structure:** Their teeth are typically fused together, forming a pair of beak-like plates in each jaw. Some species might have incisor/canine teeth as well, especially during their juvenile or initial stages.

- **Fins:** Parrotfishes possess a single continuous dorsal fin, consisting of 9 slender and flexible spines followed by 10 soft rays. The anal fin usually contains 3 flexible spines and 9 soft rays.
- **Caudal Fin:** Their caudal fin tends to be rounded to lunate in shape, often with produced caudal lobes.
- **Scales:** Parrotfishes may have 1-4 rows of scales on the cheek area and typically 2-8 median predorsal scales.

Layout

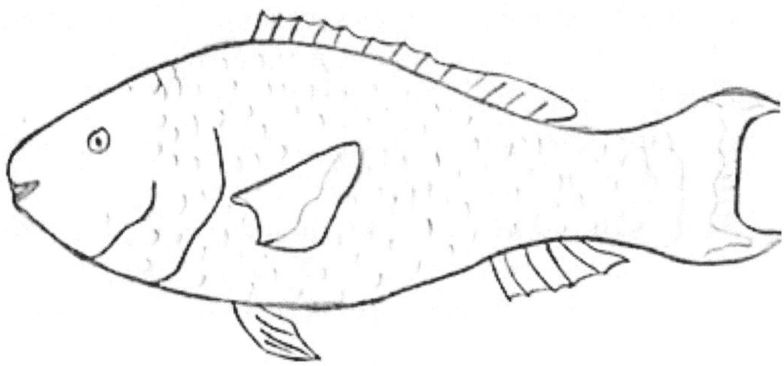

3. Sub order : Gobiodei

- Very small fishes
- Pelvic fins usually united to a cup-like structure
- Gill membranes joined to isthmus
- Dorsal spines present

Systematic classification

Kingdom: Animalia

Phylum: Chordata

Class: Actinopterygii

Order: Perciformes

Suborder: Gobiodei

Family: Gobidae.

Identification of Family: Gobidae (Gobies)

- **Morphology:** Pelvic fins fused into an adhesive disc, when well developed. Spinous dorsal present or absent; when present with 2-8 flexible spines and discontinuous with soft dorsal. Cycloid or ctenoid scales almost always present. The pelvic fins are united and forming an adhesive or sucking dics.
- First dorsal spinous and seperated from second. Five branchiostegal rays.

- They live in close association with other animals like sponges and sea urchins. This family contains world smallest fishes.
- The marine goby *Trimmatom nanus* female measures 8-10 mm. The freshwater counter part.
- *Pandaka pygmaea* and *Mistichthys luzonensis* measure upto 10-11mm. It is most species rich family of vetebrates.

Layout

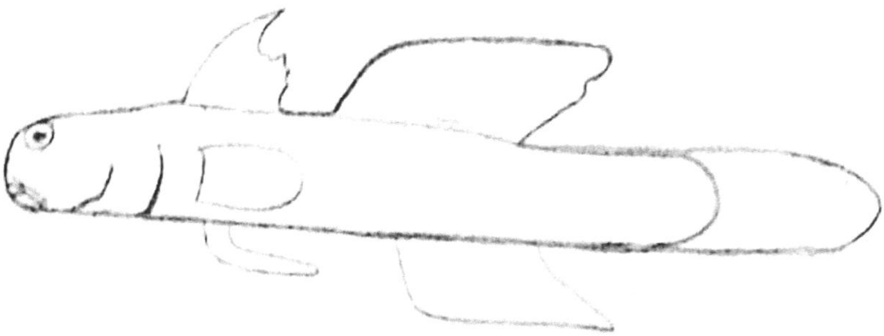

Sub order : Acanthurodei

- Small scales; one or two strong, movable forward-projecting spines
- Protrusible premaxilla
- Caudal peduncle with one spine or two bony plates

Systematic classification

Kingdom: Animalia

Phylum: Chordata

Class: Actinopterygii

Order: Perciformes

Suborder: Acanthurodei

Family: Ephippidae, Platicidae, Mugilidae, Acanthuridae, Scatophagidae, Siganidae and Zanclidae

Identification of Family: Ephippidae (Spade fishes)

- Head short, its profile either steep or fairly convex
- Pectoral fins rounded, small; pelvic fins thoracic, pointed with an axillary process
- Anal fin with 3 spines

Layout

Identification of Family: Platicidae (Bat fishes)

- Deep body, rhomboid and strongly compressed
- A single unnotched dorsal fin decreasing in size with age; spinous part not well differentiated, with 3 to 9 spines increasing in size posteriorly and nearly concealed by the scaly covering of the fin
- Short rounded, pectoral fins, pelvic fin, greatly produced.

Layout

Identification of Family: Acanthuridae (Surgeonfishes, Tangs and Unicorn fishes)

- High bodied compressed fishes with a single folding lancet-like spine or one or two bony plates usually bearing sharp keels on either side of caudal peduncle.
- Single dorsal fin.
- Caudal fin truncate to lunate.

Layout

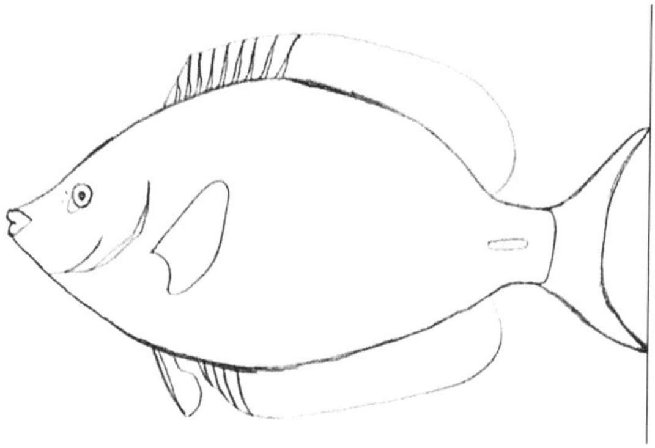

Identification of Family: Scatophagidae (Scats)

- Highly compressed body, quadrangular shaped fishes
- Head profile rising steeply to nape, mouth not protrusible
- Dorsal fin with 11 or 12 strong spines and 16 to 18 soft rays, the first spine procumbent; a deep notch between spinous and soft portions of fin
- Anal fin with 4 strong spines.

Layout

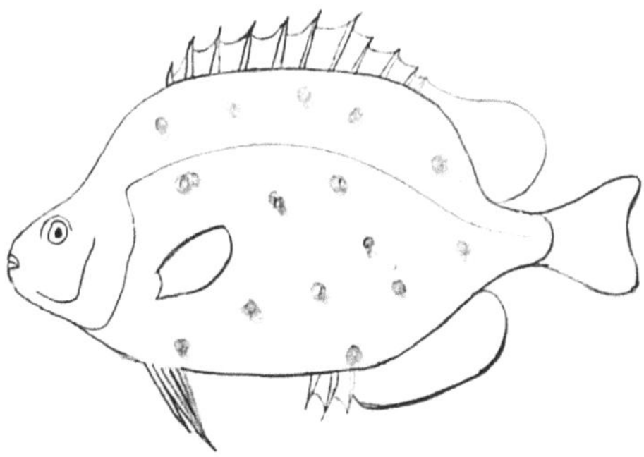

Identification of Family: Siganidae (Spine foots, Rabbit fishes)

- Body laterally compressed
- Mouth small with a single row of fine, close-set teeth in each jaw
- Dorsal fin with 13 strong spines and 10 soft rays, preceded by a forward projecting spine
- Pelvic fins with 2 strong spines separated by 3 soft rays, character unique to the family
- Anal fin with 7 strong spine and 9 soft rays
- The spines are venomous.

Layout

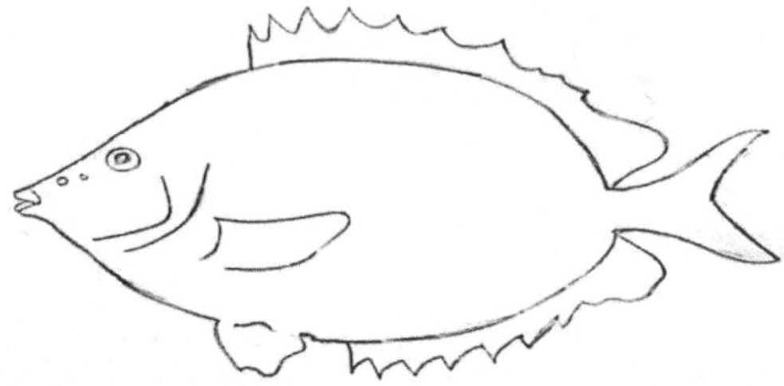

Identification of Family: Zanclidae (Moorish idol)

- Body deep and compressed, to oval or discoid body
- Tubular snout with a small mouth containing numerous elongate bristle like teeth
- Dorsal spines elongated into a whip like filament
- Small horn like projection above eye, more prominent in adults Yellow in colour with two broad blackish brown cross bands, anterior band broader crosses the eye
- Caudal dark with yellow margin and narrow blue submarginal line.

Layout

Sub order : Scombroidei

- Presence of finlets
- Keels on the caudal peduncle
- Large mouth; fast-swimming

Systematic classification

Kingdom: Animalia

Phylum: Chordata

Class: Actinopterygii

Order: Perciformes

Suborder: Scombroidei

Family: Scombridae, Sphyraenidae, Gempylidae, Trichiuridae Xiphiidae and Istiophoridae.

Identification of Family: Scombridae (Mackerals, Tunas and Bonitos)

- Body elongate and fusiform, moderately compressed in some genera.
- Snout pointed, premaxilla beak like, free from nasal bones which are separated by the ethmoid bone; mouth large; teeth in jaws strong, moderate, or weak; no true canines; palate and tongue may bear teeth. The 2 dorsal fins separate and depressible into grooves with 5-12 finlets behind second dorsal and anal fins; first dorsal fin with 9-27 rays, origin well behind the head.
- Pectoral fins high on body. Pelvic fins moderate or small with 6 fin rays, placed below the pectoral fins.
- Caudal fin deeply forked with supporting caudal rays completely covering hypural plate. At least 2 small keels on each side of caudal fin base, a larger keel in between on caudal peduncle in more advanced species.
- Lateral line simple.

Layout

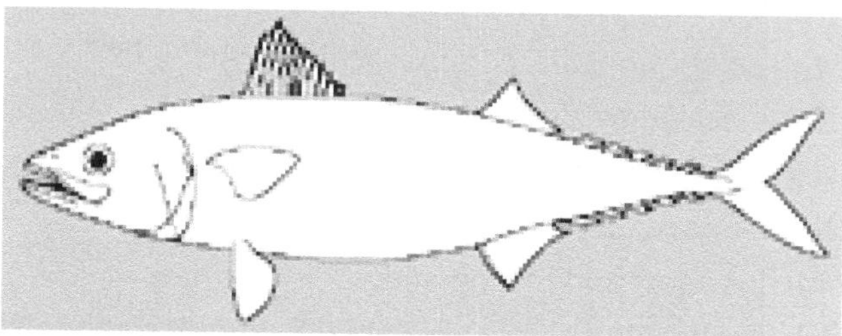

Identification of Family: Sphyraenidae (Barracudas)

- Body elongate and slightly compressed; long and pointed snout
- Lower jaw projecting beyond upper jaw; strong canine teeth
- Two short and widely separated dorsal fin, first with 5 fairly strong spines
- Short pectoral fin
- Forked caudal fin

Layout

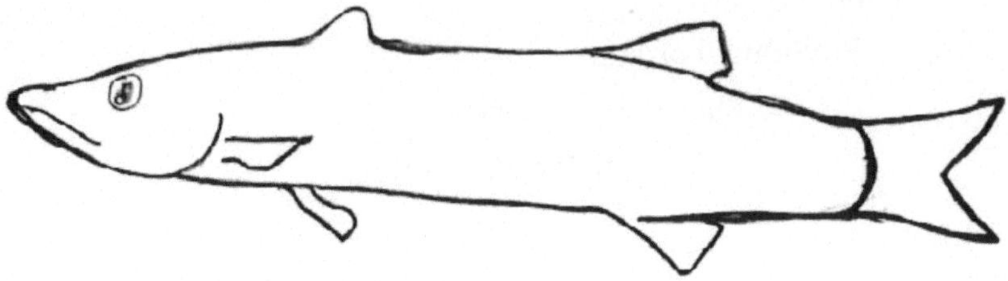

Identification of Family: Gempylidae (Snake mackerels, Barracudas)

- ❑ Lower jaw projecting beyond tip of upper jaw
- ❑ Detached finlets behind dorsal and anal fins
- ❑ Pelvic fins usually small, often reduced to a single spine with only a few or no soft rays
- ❑ No keels on caudal peduncle (except in Lepidocybium)
- ❑ Lateral line single or double, ending at caudal fin base.

Layout

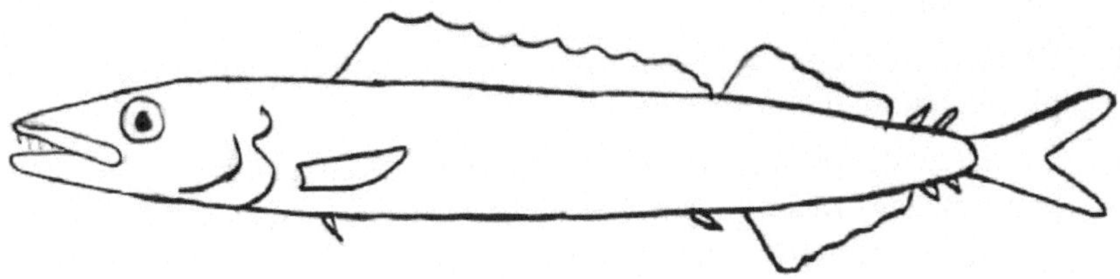

Identification of Family: Trichiuridae (Cutlassfishes / Hairtailfishes / Frostfishes / Scabbardfishes)

- ❑ Body extremely elongate, compressed and ribbon like
- ❑ Maxilla concealed by a preorbital bone
- ❑ Dorsal fin low and long, beginning shortly behind eye
- ❑ Anal fin low or reduced to short spines
- ❑ Pectoral fins short and low on body
- ❑ Pelvic fins reduced to a scale like spine (Benthodesmus sp.) or completely absent (in Trichiurus and Lepturacanthus)
- ❑ Caudal fin small, forked or absent
- ❑ Scales absent.

Layout

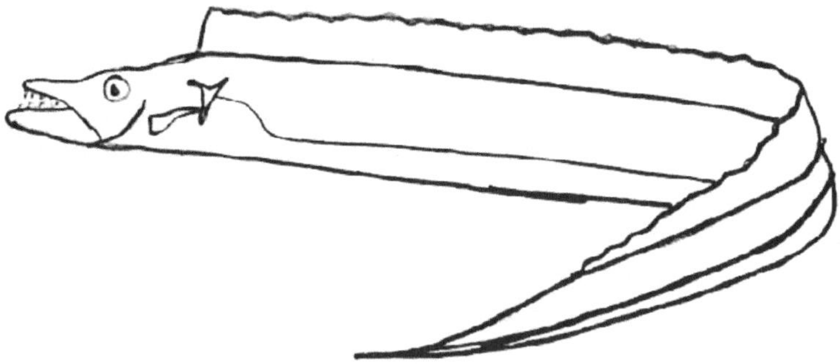

Identification of Family: Xiphiidae (Swordfishes)

- ❑ Body elongate and cylindrical
- ❑ Upper jaw prolonged into a long bill (but both jaws prolonged into long bills in immature individuals)
- ❑ No gill rakers on gill arches
- ❑ Two separate dorsal fin in adults (continuous in immature specimens) the first much larger than the second
- ❑ Pectoral fins falcate
- ❑ Pelvic fins absent
- ❑ A large keel on each side of caudal peduncle
- ❑ Lateral line indistinct, but recognizable in specimens of 1 m length.

Layout

Identification of Family: Istiophoridae (Bill fishes, Spearfishes, Marlins and Sailfishes)

- ❑ Upper jaw prolonged into a long spear which is round in cross section
- ❑ No rakers on gill arches
- ❑ Two dorsal fins and two anal fins
- ❑ 1 st dorsal fin and anal fin long, can both be folded back into grooves.
- ❑ Pectoral fins falcate

- Caudal fin large, strong and forked with a pair of keels on either side at base
- Pelvic fin very long with 1 or 2 rays and a spine.

Layout

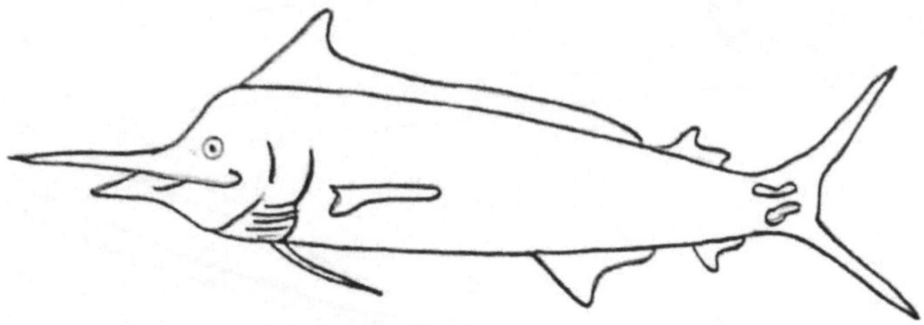

Sub order : Stromateoidei

- Variable shapes
- Blunt, thick snout
- Small mouth; sometimes pelvic fin absent
- Toothed saccular outgrowths in gullet behind last gill arch
- Caudal fin with 15 branched rays

Systematic classification

Kingdom: Animalia

Phylum: Chordata

Class: Actinopterygii

Order: Perciformes

Suborder: Stromateoidei

Family: Stromateidae

Identification of Family: Stromateidae (Butterfishes and Silver pomfrets)

- Body very deep and compressed
- Snout short and blunt; mouth small and immobile Maxilla
- Single dorsal fin and anal fin; long-based and slightly to deeply falcate, preceeded by none or 5 to 10 flat, blade-like spines.
- Pectoral fins long and wing-like
- No pelvic fins.
- Caudal fin forked; in some species with very extended lobes
- Gill membranes broadly united to the isthmus.

Layout

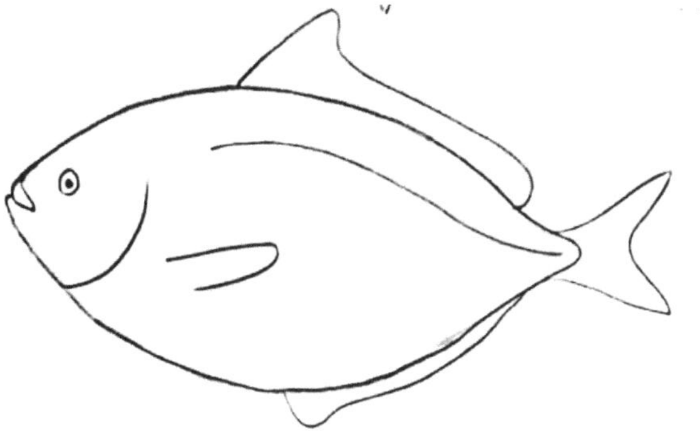

INFERENCE:

Chapter-17

Identification of Order – Pleuronectiformes

Introduction

A flatfish is a member of the ray-finned demersal fish order Pleuronectiformes, also called the Heterosomata, sometimes classified as a suborder of Perciformes. In many species, both eyes lie on one side of the head, one or the other migrating through or around the head during development.

Aim

The aim of a practical exercise is focused on the identification of the order: Pleuronectiforms.

Systematic Classification

Kingdom: Animalia

Phylum: Chordata

Class: Actinopterygii

Order: Pleuronectiformes

Family: Psettodidae, Citharidae, Pleuronectidae, Soleidae and Cynoglossidae and Bothidae

Identification Key Characters of Family – Bothidae

- ❑ Eyes present in the left side of the body

- Preopercle margin free and distinct.
- No spines in fins.
- Dorsal fin origin above or anterior to upper eye.
- Dorsal and anal fins separate from caudal fin.
- Branchiostegal membranes united.
- Anus on blind side.

Layout

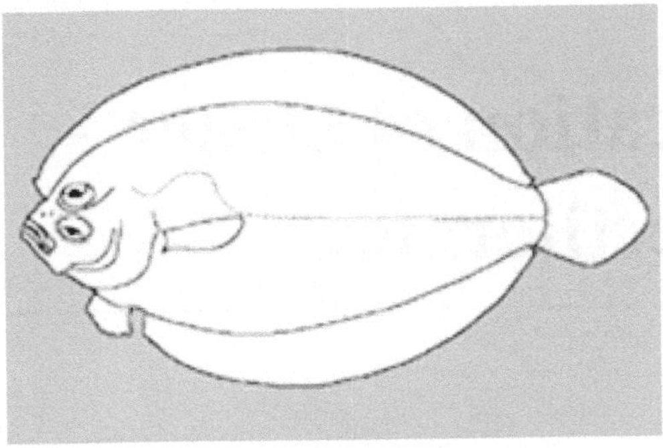

Identification Key Characters of Family – Citharidae

- Eyes present in the left or right side of the body
- A single spine and five soft rays in pelvic fins; pelvic equally developed and short based.
- Dorsal fin origin at posterior nostril of blind side; pectoral Fins well developed.
- Branchiostegal membranes not united.
- Lateral line present on both sides.
- Anus on eyed side.

Layout

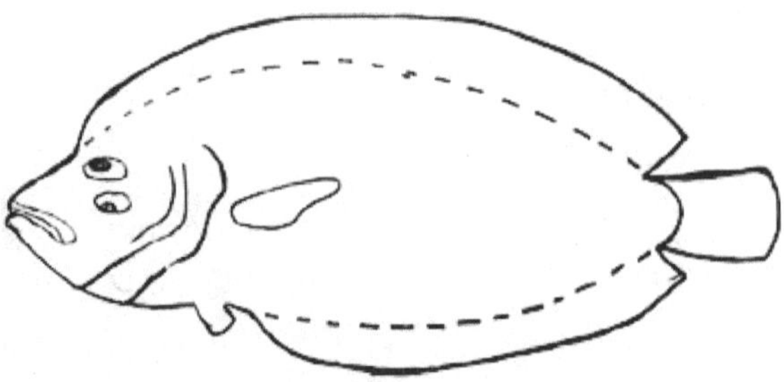

Identification Key Characters of Family – Cynoglossidae

- Eyes very small with little interorbital space.

- Eyes present in the left side of the body
- Preopercle margin covered with skin.
- Asymmetrical mouth.
- Caudal fin pointed and confluent with dorsal and caudal fins.
- Right pelvic fin usually undeveloped.
- No pectoral fins.

Layout

Identification Key Characters of Family – Pleuronectidae

- Eyes present in the right side of the body
- Preopercular with a free margin.
- Fins without spines.
- Dorsal fin extends on head.
- Swim bladders absent in adults.

Layout

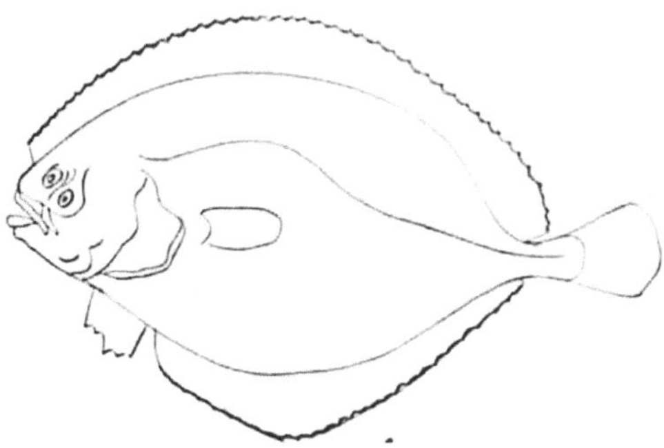

Identification Key Characters of Family – Psettodidae

- Large mouth; vomer, palatine, and tongue with minute teeth.
- Distinct preopercular margin, not hidden beneath skin.
- Eyes present in the right or left side of the body
- Dorsal fin origin well behind eyes, the anterior rays spinous.

- Almost symmetrical pelvic fins with a weak spine and five soft rays.
- Cycloid scale: lateral line developed on both sides of body.

Layout

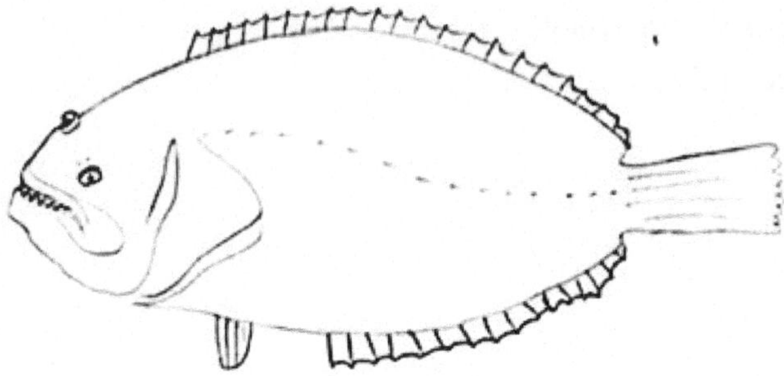

Identification Key Characters of Family – Soleidae

- Margin of pre operculum completely concealed (covered by skin and scales).
- Eyes present in the right of the body
- Dorsal fin origin above or before eyes; pectoral fins present or absent; no fin spines.
- Dorsal and anal fins free from caudal fin or united with caudal fin.
- Pelvics free from anal fin.

Layout

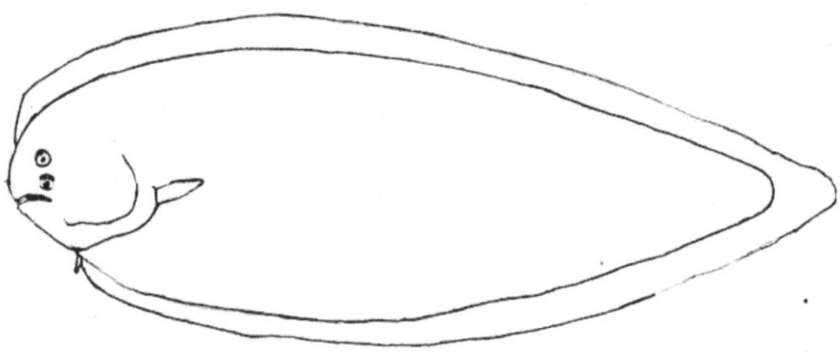

INFERENCE:

Chapter-18

Identification of Order – Siluriformes (Marine Water)

Introduction

Siluriformes, an order of fish, exhibit distinct morphological characteristics such as reduced mesopterygoid bone and relatively small opercle and interopercle. They often possess an adipose fin and spine-like rays on their dorsal and pectoral fins. Their fins and skeletal features vary, with dorsal fins typically sporting two spines and caudal fins having 18 or fewer rays. External traits include naked or bony-plated bodies, up to four pairs of barbells on the head, and usually small eyes. Adaptations like air-breathing organs are seen in certain families, and size variations range from small to exceptionally large species like those in the Pangasiidae Pimelodidae families. Siluriformes are distributed across various habitats, from marine to freshwater, with some species capable of brackish water living. Commercially significant species in marine environments include Ariidae and Plotosidae.

Aim

The aim of a practical exercise is focused on the identification of the order: Siluriformes.

Systematic Classification

Kingdom: Animalia

Phylum: Chordata

Class: Actinopterygii

Order: Siluriformes

Family: Ariidae and Plotosidae.

Identification Key Characters of Family – Ariidae

- ❑ Snout and head rounded to depressed
- ❑ 1-3 pairs of barbels present
- ❑ Head covered with a bony shield
- ❑ A short adipose dorsal fin present
- ❑ First dorsal fin short with a short spine or buckler.
- ❑ Two pairs of adjacent nostrils on each side of snout.

Layout

Identification Key Characters of Family – Plotosidae

- ❑ Elongate body, compressed, tapering to a point posteriorly
- ❑ 4 pairs of barbels present with 1 pair nasal, 1 pair maxillary and 2 pairs mental;
- ❑ First dorsal fin short-based with a serrated spine and 4-6 soft rays; second dorsal fin (or dorsal procurrent caudal fin), caudal fin and anal fin confluent
- ❑ Absence of adipose fin.

Layout

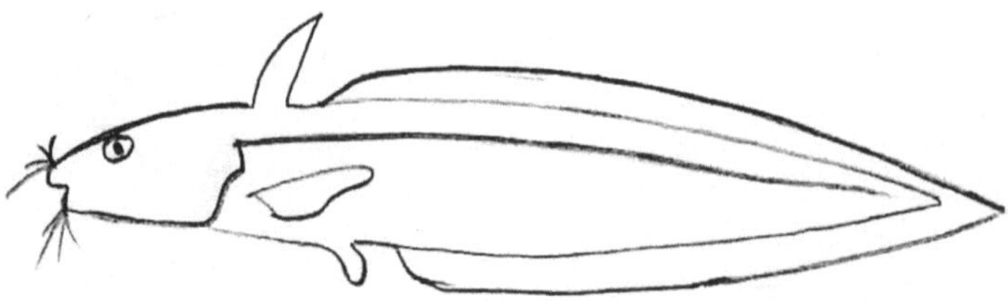

Chapter-19

Identification of Order – Gonorynchiformes

Introduction

The Gonorynchiformes are an order of ray-finned fish that includes the important food source, the milkfish, and a number of lesser-known types, both marine and freshwater.

Aim

The aim of a practical exercise is focused on the identification of the order: Gonorynchiformes.

Systematic Classification

Kingdom: Animalia

Phylum: Chordata

Class: Actinopterygii

Order: Gonorynchiformes

Family: Chanidae

Identification Key Characters of Family – Chanidae (Milk fish)

- **Body Shape:** Fusiform (elongated and tapering at both ends) and compressed.
- **Mouth Position:** Terminal (at the end of the head).
- **Jaw Structure:** Upper jaw is non protrusible.

- **Scales:** Cycloid (smooth-edged scales).
- Pelvic fin with 11 or 12 rays.
- Dorsal fin with 13-17 rays.
- Anal fin with 9-11 rays.
- Four branchiostegal rays.
- **Swim Bladder:** Present.

Layout

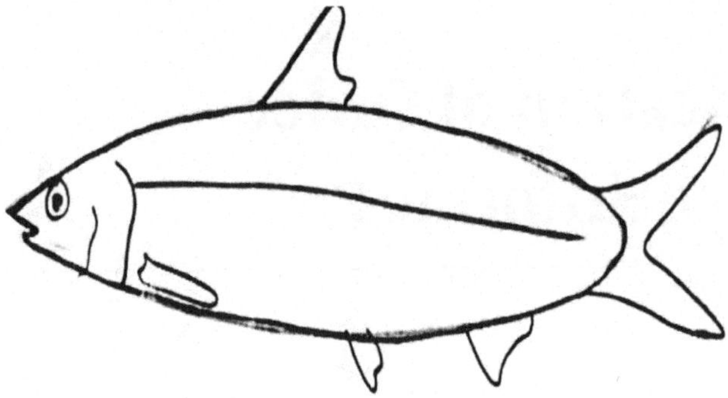

INFERENCE:

Chapter-20

Identification of Order – Berciformes

Introduction

The Berciformes are a poorly understood order of carnivorous ray-finned fishes consisting of 7 families, 30 genera, and 161 species. They feed on small fish and invertebrates.

Aim

The aim of a practical exercise is focused on the identification of the order: Berciformes.

Systematic Classification

Kingdom: Animalia

Phylum: Chordata

Class: Actinopterygii

Order: Berciformes

Family: Holocentridae

Identification Key Characters of Family – Holocentridae (Squirrel fishes, Soldier fishes)

- **Pelvic Fin Spine:** Typically, a single spine exists with 5-8 soft rays (most commonly 7) in the pelvic fin.

- **Dorsal Fin:** Long dorsal fin comprising 10-13 spines and a notched soft-rayed section with 11-17 rays.
- **Anal Fin:** Anal fin with 4 spines, the third the stoutest and often the longest with Consists of four spines and 7-16 soft rays.
- **Caudal Fin:** Forked caudal fin with principal caudal rays numbering 18 or 19.
- **Scales:** Exhibits large and rough ctenoid scales.
- **Eyes:** Large eye structure.
- **Operculum:** Possesses a spiny-edged operculum.
- **Vertebrae:** Typically has 26 or 27 vertebrae.
- **Coloration:** Usually reddish in colour.

Layout

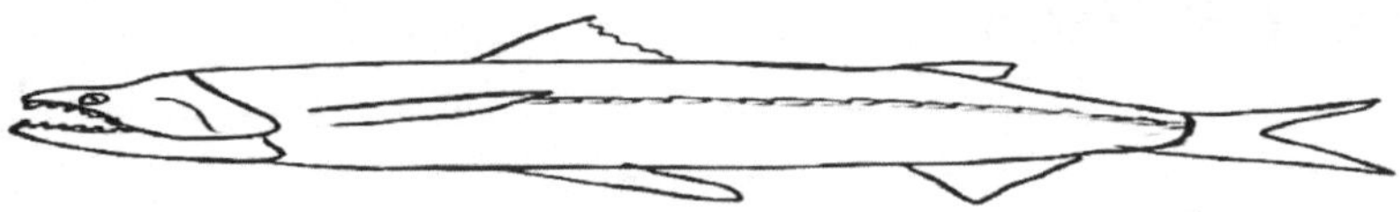

INFERENCE:

Chapter-21

Identification of Order – Aulopiformes

Introduction

Aulopiformes is a diverse order of marine ray-finned fish consisting of some 15 extant and several prehistoric families with about 45 genera and over 230 species. The common names grinners, lizardfishes and allies, or aulopiforms are sometimes used for this group. **Aim**

The aim of a practical exercise is focused on the identification of the order: Aulopiforms.

Systematic Classification

Kingdom: Animalia

Phylum: Chordata

Class: Actinopterygii

Order: Aulopiformes

Family: Harpadontidae and Synodontidae

Identification Key Characters of Family – Harpadontidae (Bombay duck)

- **Dorsal Fin with Adipose Fin:** A prominent dorsal fin followed by a noticeable adipose fin.

- **Pelvic Fins:** Remarkably long pelvic fins.
- **Lateral Line Extension:** The lateral line extends into a pointed median lobe of the caudal fin.

Layout

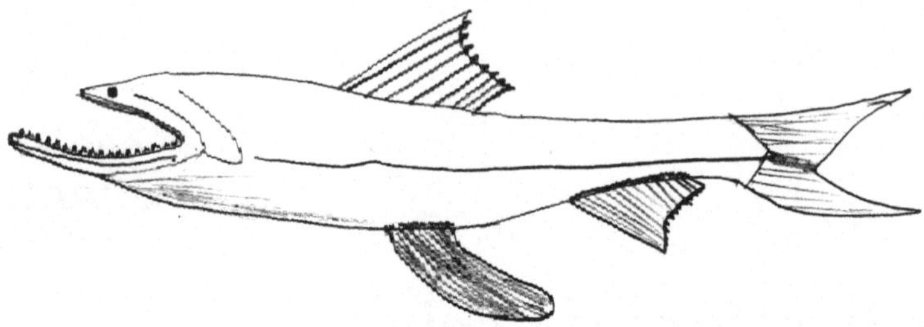

Identification Key Characters of Family – Synodontidae (Lizard fish)

- Small, cylindrical body structure.
- Fins lack spines.
- Large mouths equipped with slender, sharp teeth, including those on the tongue.
- Elongate and cylindrical body with adipose fin.
- Pointed teeth visible even when mouth is closed.

Layout

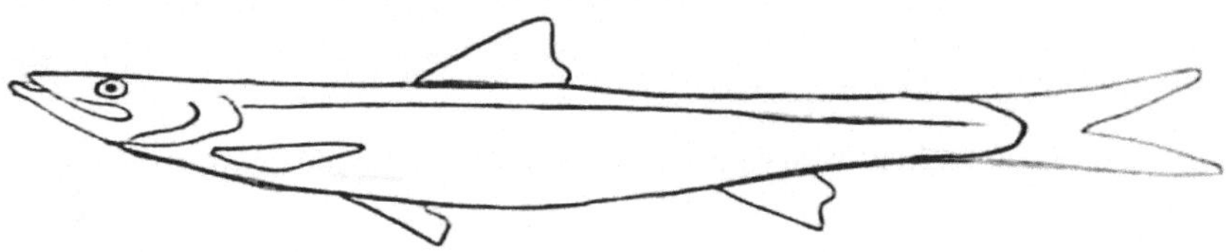

INFERENCE:

Chapter-22

Identification of Order – Elopiformes

Introduction

The Elopiformes are the order of ray-finned fish including the tarpons, tenpounders, and ladyfish, as well as a number of extinct types. They have a long fossil record, easily distinguished from other fishes by the presence of an additional set of bones in the throat .

Aim

The aim of a practical exercise focused on the identification of the order: Elopiformes.

Systematic Classification

Kingdom: Animalia

Phylum: Chordata

Class: Actinopterygii

Order: Elopiformes

Family: Elopidae and Megalopidae

Identification Key Characters of Family – Elopidae (Tenpounders)

- Well-developed gular plates between the lower jaws

- Single dorsal fin with origin slightly behind the midpoint of the body

Layout

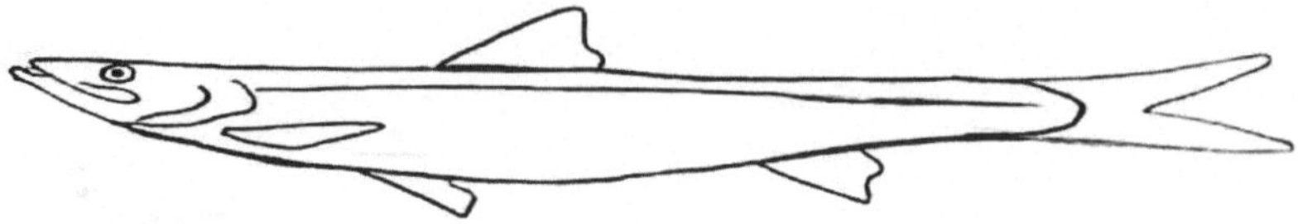

Identification Key Characters of Family – Megalopidae (Tarpons)

- Large eye
- Gular plate located between arms of lower jaw
- Single dorsal fin present at the midpoint of the body; last dorsal fin ray filamentous.

Layout

INFERENCE:

Chapter-23

Identification of Order – Beloniformes

Introduction

Beloniformes are long, silvery streamlined fishes that live near the surface and feed on plankton, small fishes and invertebrates. Species within the group have a variety of jaw lengths and shapes, and most species go through a half-beak stage during development, where the lower jaw is elongated. Most have a counter-shading colour pattern to avoid being seen by prey and predators. They are greenish to bluish on the back and silvery white on the sides and belly, sometimes with a dusky or darker stripe along the sides. Most species are marine, although some live in brackish and freswater areas. The order comprises five families and four of these are found in Australian waters: needlefishes, flying fishes, halfbeaks and sauries. The flesh is excellent, and the group contains a number of commercially important species. Some species are regularly traded in the aquarium industry. Beloniformes means "needle-shaped" from the Greek, belone meaning needle, and the Latin forma meaning shape. The fossil record for the group dates back to the early Tertiary Period.

Aim

The aim of a practical exercise is focused on the identification of the order: Beloniformes.

Systematic Classification

Kingdom: Animalia

Phylum: Chordata

Class: Actinopterygii

Order: Beloniformes

Family: Belonidae, hemiramphidae and Exocoetidae

Identification Key Characters of Family – Belonidae (Needle fishes)

- Elongate fishes with both upper and lower jaws extended into a long beak
- Dorsal and anal fins posterior in position
- Pelvic fins located in abdominal position.

Layout

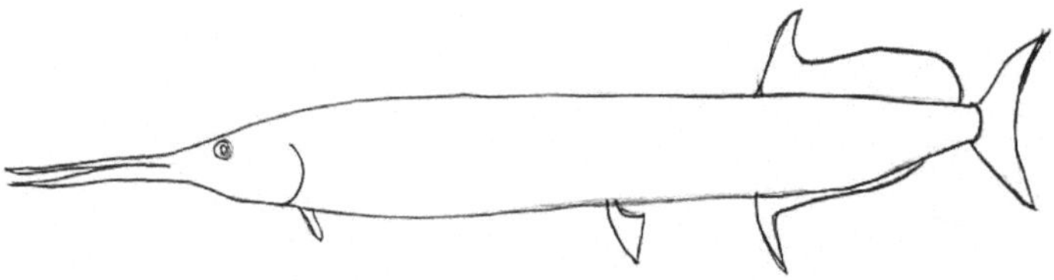

Identification Key Characters of Family – Hemiramphidae (Halfbeaks)

- Elongate fishes with a prolonged lower jaw (except Oxyporhampus) and a short triangular upper jaw
- Dorsal and anal fins posterior in position; pelvic fins in abdominal position
- Lateral line running down from pectoral fin origin and then backward along the ventral margin of the body.

Layout

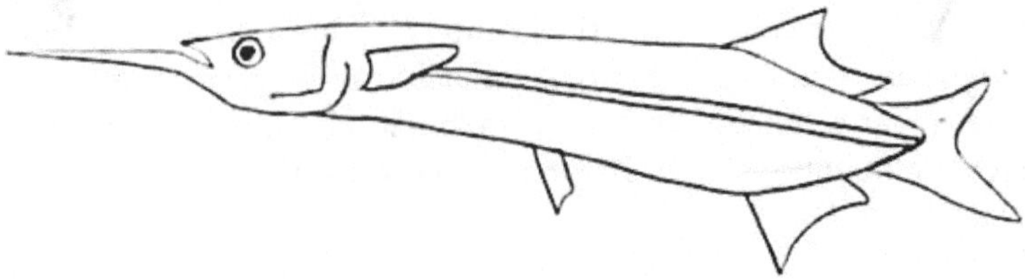

Identification Key Characters of Family – Exocoetidae (Flying fish)

- Pectoral fins high on sides, strikingly long, always extending beyond dorsal fin origin
- Pelvic fins abdominal in position and greatly enlarged in many
- Caudal fin deeply forked with lower lobe longer than the upper.

Layout

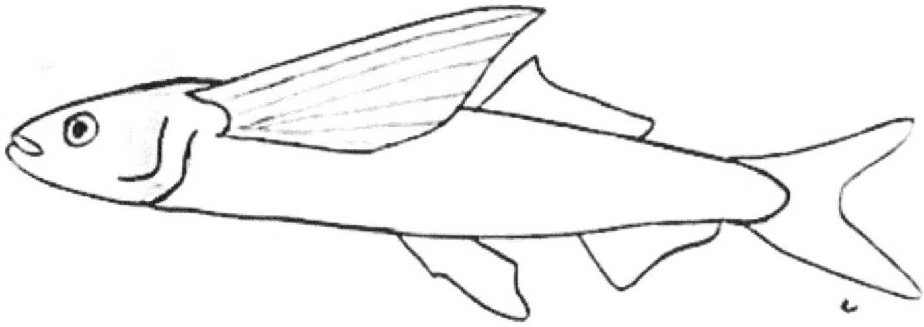

INFERENCE:

Chapter-24

Identification of Order – Mugiliformes

Introduction

The mullets or grey mullets are a family of ray-finned fish found worldwide in coastal temperate and tropical waters, and some species in fresh water. Mullets have served as an important source of food in Mediterranean Europe since Roman times. The family includes about 78 species in 20 genera.

Aim

The aim of a practical exercise focused on the identification of the order: Mugiliformes.

Systematic Classification

Kingdom: Animalia

Phylum: Chordata

Class: Actinopterygii

Order: Mugiliformes

Family: Mugilidae

Identification Key Characters of Family – Mugilidae (Grey Mullet)

- **Adipose Eyelid:** Presence of an adipose eyelid.

- **Head Structure:** Flattened head, contributing to distinctive morphology.
- **Dorsal Fins:** Usually equipped with two dorsal fins.
- **Body Coloration:** Silvery body, providing a characteristic appearance.
- **Lateral Line:** Not present.
- **Body Shape:** Elongated body structure.
- **Mouth Size:** Relatively small mouths
- **Pectoral Fin Placement:** Pectoral fin positioned high on the body.
- Travel in schools and feed on fine algae, diatoms, and detritus of bottom sediments. Important food fishes.

Layout

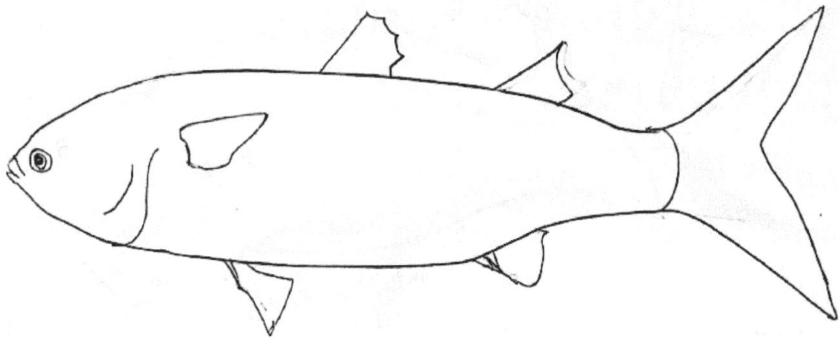

INFERENCE:

Chapter-25

Identification of Order – Tetraodontiformes

Introduction

Tetraodontiforms are a diverse group of tropical marine fish related to perciformes, evolving about 50 million years ago. This order includes triggerfishes, puffers, filefishes, and porcupine fishes, totaling around 360 species. They exhibit a wide range of anatomical structures and behaviors. While some families like Triacanthodidae show a spectrum of configurations from normal to highly specialized forms, others like Balistidae are more uniform. Monacanthids encompass species with varying body plans from normal to extremely elongated and specialized.

Aim

The aim of a practical exercise is focused on the identification of the order: Tetraodontiformes.

Systematic Classification

Kingdom: Animalia

Phylum: Chordata

Class: Actinopterygii

Order: Tetraodontiformes

Family: Tricanthidae, Balistidae, monacanthidae, Ostraciidae, Tetraodontidae and Diodontidae

Identification Key Characters of Family: Triacanthidae (Tripod fishes, Triplespines)

- Deep Compression body.
- Shagreen-Like Skin Appearance
- Dorsal Fin commonly having 5 Spines, Rudimentary 6th Spine
- Deeply Forked Caudal Fin
- Tapered Caudal Peduncle
- Large Spine as Pelvic Fin

Layout

Identification Key Characters of Family: Balistidae (Trigger fishes and Durgons)

- Deep and Moderately Compressed Body
- Very Thick, Tough Skin
- Tympanum Formed by Scales Above Pectoral Fin Base
- Two Dorsal Fins
- First Dorsal Fin with 3 Spines
- Second Dorsal Fin with Branched Fin-Rays
- Relationship Between First and Second Dorsal Fin Spines
- Locking Mechanism of First Spine
- Rudimentary or Absent Pelvic Fins
- Enlarged Scales Encasing Pelvic Region.

Layout

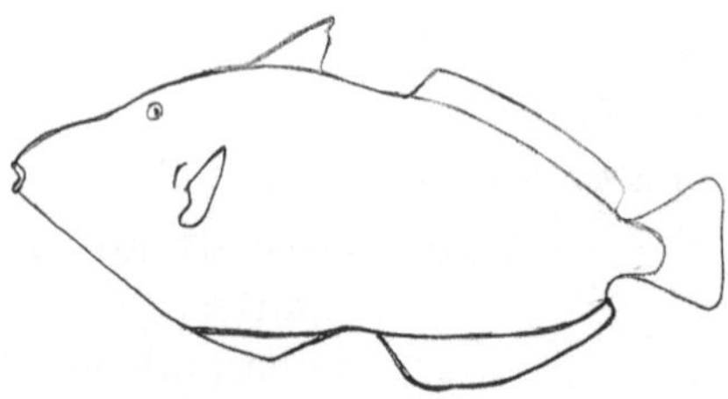

Identification Key Characters of Family: Monacanthidae (Filefishes / Leather jackets)

- ❑ Smooth to Rough Shagreen-Like Skin
- ❑ Minute Scales with Fine Spicules
- ❑ Two Dorsal Fin Spines
- ❑ Large First Spine with Downward-Curving Barbs
- ❑ Locking Mechanism of the First Spine with the Small Second Spine
- ❑ Rudimentary Pelvic Fins
- ❑ Enlarged Scales Encasing the Pelvic Region
- ❑ Variability in the Presence of Pelvic Fins with Age.

Layout

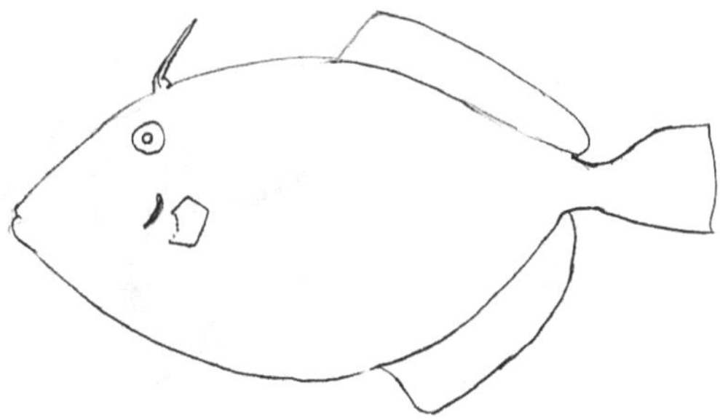

Identification Key Characters of Family: Ostraciidae (Box fishes/ Cow fishes)

- ❑ Body almost completely encased in a bony shell, or carapace formed of enlarged, thickened scale plates
- ❑ Absence of pelvic fin.
- ❑ Dorsal fin rays 9-13. Anal fin rays 9-13.
- ❑ Non-protrusible upper jaw.

Layout

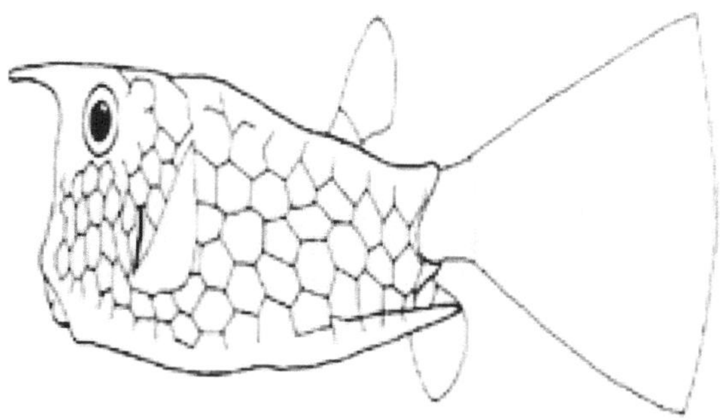

Identification Key Characters of Family: Tetraodontidae (Puffer fishes/ Blow fishes/ Tobies)

- ❑ Heavy and Blunt Body
- ❑ Rapid Inflation Mechanism through Water or Air Intake
- ❑ Large and Blunt Head
- ❑ Jaws Modified to Form a Beak
- ❑ Heavy, Powerful Teeth (2 Above and 2 Below)
- ❑ Eyes Located High on Head
- ❑ Absence of Pelvic Fins
- ❑ Dorsal and Anal Fins Located Far Posteriorly
- ❑ No Dorsal Fin or Anal Fin Spines
- ❑ Absence of Typical Scales
- ❑ Numerous Small Spiny Prickles on Back and/or Belly
- ❑ Occasional Presence of Prickles on Sides.

Layout

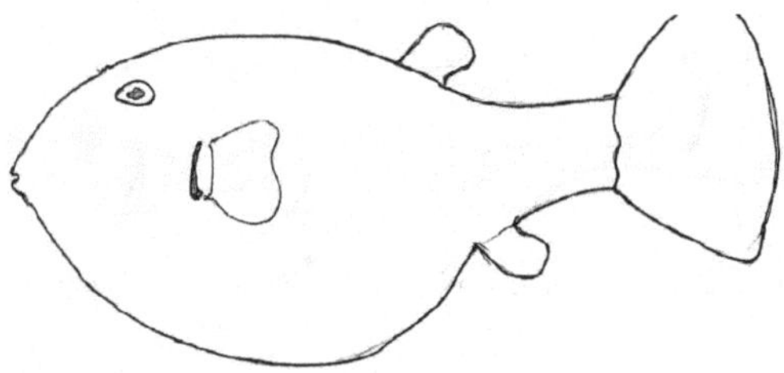

Identification Key Characters of Family: Diodontidae (Porcupine fishes, Spiny puffers, Baloon fishes)

- ❑ Wide Body Capable of Great Inflation
- ❑ Presence of Spines Covering the Body skins.
- ❑ Teeth Fused to Form Beak-Like Crushing Structure
- ❑ Division of Upper and Lower Jaws into Left and Right Halves
- ❑ Lack of Pelvic Fins
- ❑ Inconspicuous Lateral Line
- ❑ Absence of Normal Scales.

Layout

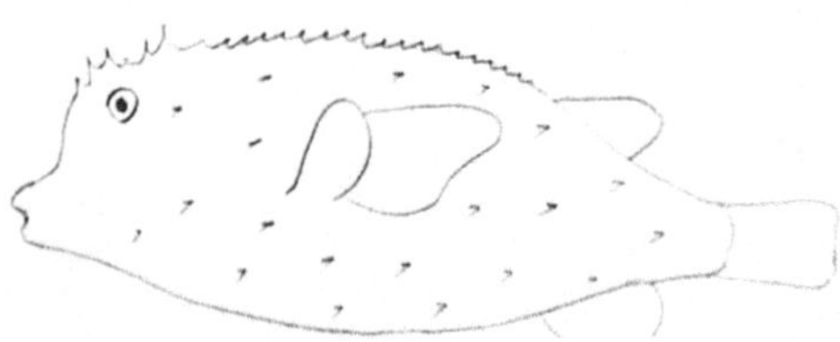

INFERENCE:

Chapter-26

Identification of Order – Anguilliformes

Introduction

Eels, members of the order Anguilliformes, represent a diverse group of ray-finned fish encompassing eight suborders, 20 families, approximately 164 genera, and boasting around 1000 species. Their life cycle is marked by significant developmental changes from the early larval stage to adulthood. Typically, eels are predators, exhibiting a wide array of hunting strategies across their various habitats, from freshwater streams to deep-sea environments

Aim

The aim of a practical exercise is focused on the identification of the order: Anguilliformes.

Systematic Classification

Kingdom: Animalia

Phylum: Chordata

Class: Actinopterygii

Order: Anguilliformes

Family: Anguillidae, Muraenidae, Ophichthidae, Muraenesocidae, Nemichthyidae, Congridae and Xenocongridae.

Identification Key Characters of Family: Anguillidae (Freshwater eels)

- ❑ Small Oval Scales Embedded in Skin
- ❑ Basket-Weave Pattern of Scale Arrangement
- ❑ Elongate and Snake-Like Body
- ❑ Lower Jaw Longer than Upper
- ❑ Projecting Mouth, Angle behind Near Margin of Eye
- ❑ Absence of Pelvic Fin
- ❑ Well-Developed Pectoral Fins
- ❑ Anus in Anterior Half of the Body
- ❑ Vertical Slit-Like Gill Opening
- ❑ Complete Lateral Line on Body and Head
- ❑ Vertebrae Count (100s – 119).
- ❑ Catadromous and Spawning Migration
- ❑ Leptocephali Metamorphosis into Elvers.

Layout

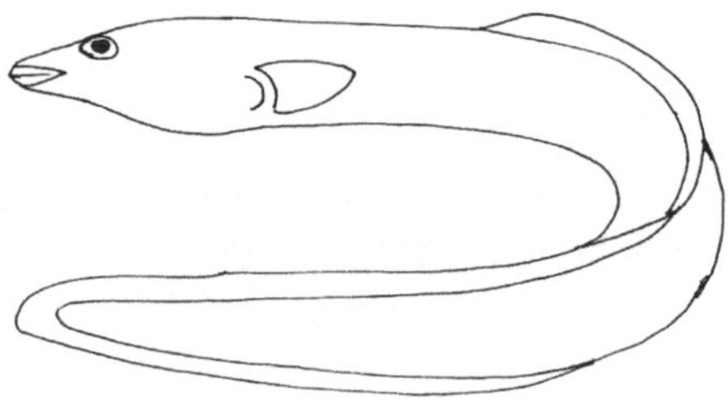

Identification Key Characters of Family: Muraenidae (Morray eel)

- ❑ **Dorsal Profile**: Steep above and behind the eye.
- ❑ **Gill Opening**: Small, roundish lateral hole or slit.
- ❑ Dorsal fin and anal fin are continuous around the tail. Pectoral and pelvic fins are absent.
- ❑ **Lateral Line Pores:** No lateral line pores on the body, but a reduced complement on the head, including 1 or 2 above and before the gill opening (branchio lateral line pores).
- ❑ **Scales:** No scales.
- ❑ **Gill Arch:** Reduced gill arches.
- ❑ **Nostril Position:** Posterior nostril is high on the head.
- ❑ **Teeth:** Fang-like teeth are present.
- ❑ **Vertebrate Count:** Usually 110-200.
- ❑ **Genera:** Over 200 species reported under 15 genera.

Layout

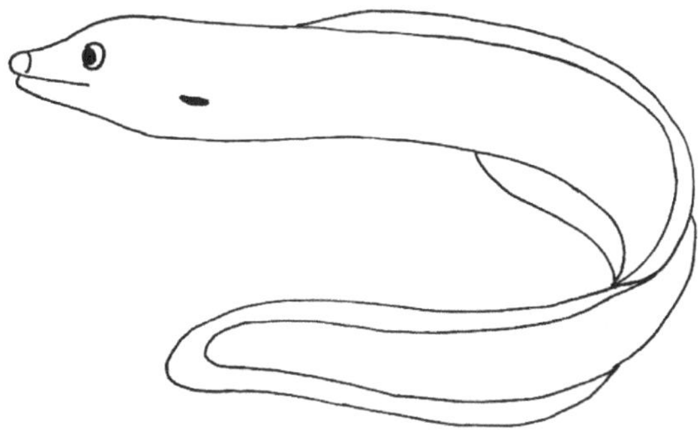

Identification Key Characters of Family: Ophichthidae (Worm eels and snake eels)

- The characteristics described align with the family Anguillidae, commonly known as freshwater eels.
- **Body Shape:** Elongate, slender, and cylindrical.
- **Gill Opening:** Small.
- **Pectoral Fins:** Absent.
- **Dorsal Fin:** Originating about a pectoral fin length behind the tips of the pectoral fins.
- **Vertebrate Count**: Typically 171-173.

Layout

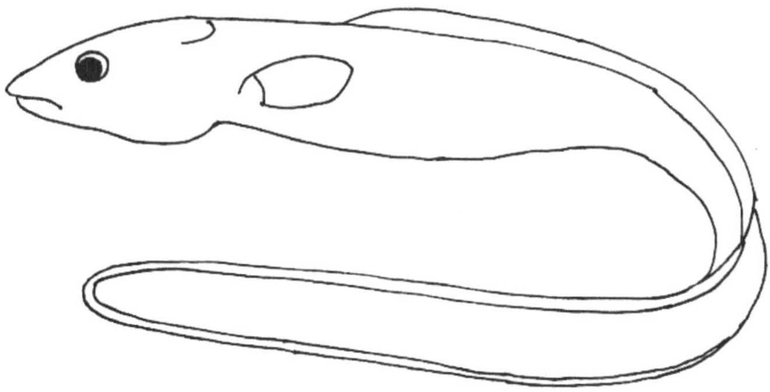

Identification Key Characters of Family: Muraenesocidae (pike congers)

- **Long and Cylindrical:** The body is long to very long, more or less cylindrical in front.
- **Compressed Tail:** The body is compressed along the tail.
- **Well-developed Mouth:** The mouth is well developed and extends beyond the eye.

- **Terminal Mouth:** The mouth is terminal, extending well beyond the eye.
- **Dorsal Fin:** The dorsal fin begins more or less above the gill opening, and the snout is very pointed.
- **No Pelvic Fin:** There is no pelvic fin.
- **Pectoral Fins:** Pectoral fins are present.
- **No Scales:** The species lacks scales.
- **Well-developed Teeth:** Teeth are well developed and fang-like.
- Eyes large and covered with skin
- Body very long with only tail compressed
- Snout elongate
- Vertebrae 120-216

Layout

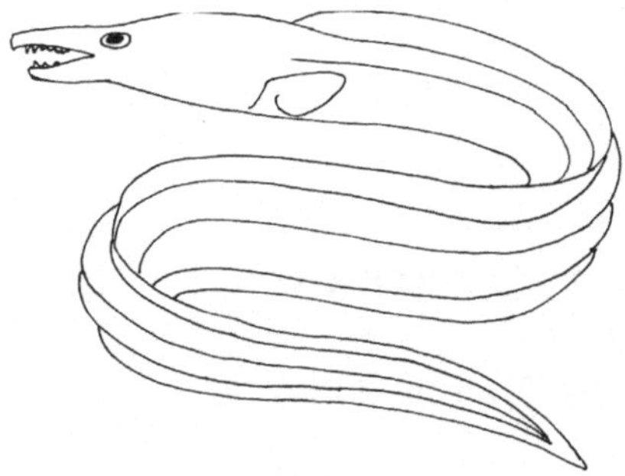

Identification Key Characters of Family: Nemichthyidae (Snipe eels)

- **Extremely Long Jaws:** Jaws are extremely long, needle-like.
- **Body Shape:** The body is long and slender.
- **Pectoral Fin:** Pectoral fin is present.
- **Large Eyes:** Eyes are relatively large.
- **Anus Location:** The anus is positioned far forward.

Layout

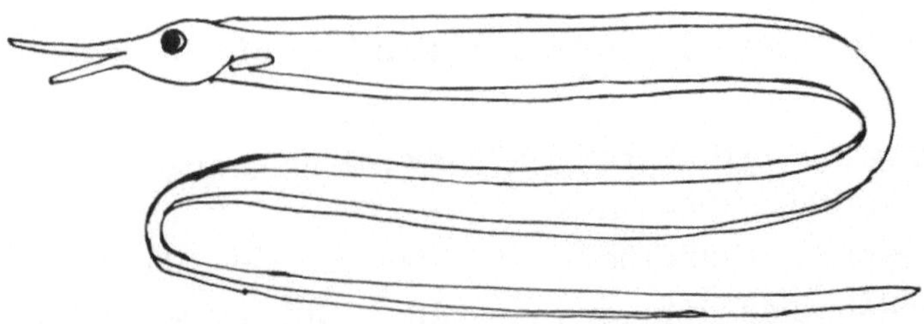

Identification Key Characters of Family: Congridae (Conger eels)

- ❑ **Dorsal and Anal Fins:** Dorsal fin and anal fin are continuous around the tail.
- ❑ **Dorsal Fin Origin:** Dorsal fin begins more or less above the gill opening and originates before pectoral fin tip.
- ❑ **No Pelvic Fins:** There are no pelvic fins.
- ❑ **Body Structure:** The body is robust without scales.
- ❑ **Supratemporal Pore:** Supratemporal pore is present in front of dorsal fin.
- ❑ **Lateral Line:** Lateral line is complete.
- ❑ **Mouth Shape:** Mouth does not extend beyond the eye.
- ❑ **Well-developed Teeth:** Teeth are well developed, but no canines in the jaw.
- ❑ **Jaw Length:** Lower jaw is equal to or shorter than the upper jaw.
- ❑ **Vertebrae Count:** The number of vertebrae ranges from 205 to 225.

Layout

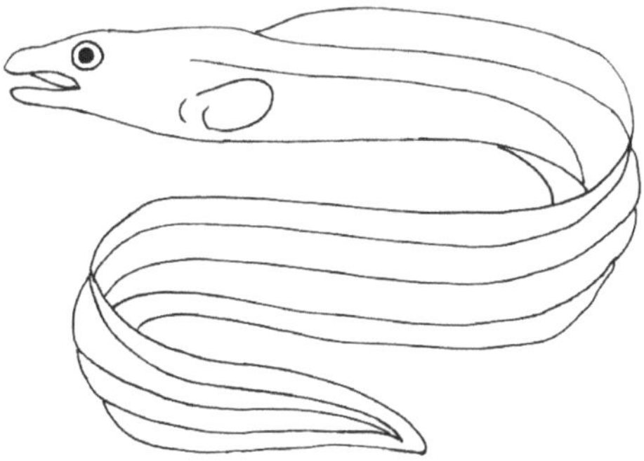

INFERENCE:

Chapter-27

Identification of Order – Syngnathiformes

Introduction

A group of highly modified and specialised ray finned fishes which include Seahorses, Pipe fishes, Sea dragons, Ghost pipe fishes, etc. The order comprises 6 families and more than 370 species.

Aim

The aim of a practical exercise focused on the identification of the order: Syngnathiformes

Systematic Classification

Kingdom: Animalia

Phylum: Chordata

Class: Actinopterygii

Order: Syngnathiformes

Family: Sygnathidae, Solenoceridae, Countryside, Fistulariidae, Pegasidae and Aulostomidae.

Identification Key Characters of Family: Syngnathidae(Pipe fishes)

- Joined mouth
- Head extended, separated from body
- Body encased in numerous bony plates
- Dorsal fin present
- Anal fin short (2- rays)
- Pelvic fin absent
- CF absent.
- Shallow water found

Layout

Identification Key Characters of Family: Solenostomidae (Ghost pipe fish / False pipe fish)

- Long Pelvic fin
- Body relatively short
- Dorsal fin - two
- First dorsal 5 spine
- Gill opening is moderately large
- Brood pouch is found in female

Layout

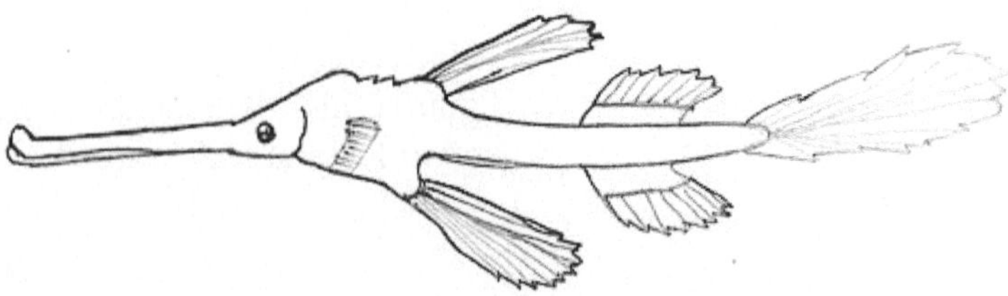

Identification Key Characters of Family: Centriscidae(Razor fish)

- ❑ Found like thin razor
- ❑ Body compressed laterally
- ❑ First dorsal with single spine
- ❑ Body fully covered with thin body plates
- ❑ Caudal and second dorsal displaced ventrally.
- ❑ Mouth Toothless
- ❑ No lateral line and usually swim in vertical position
- ❑ Shrimp fishes.

Layout

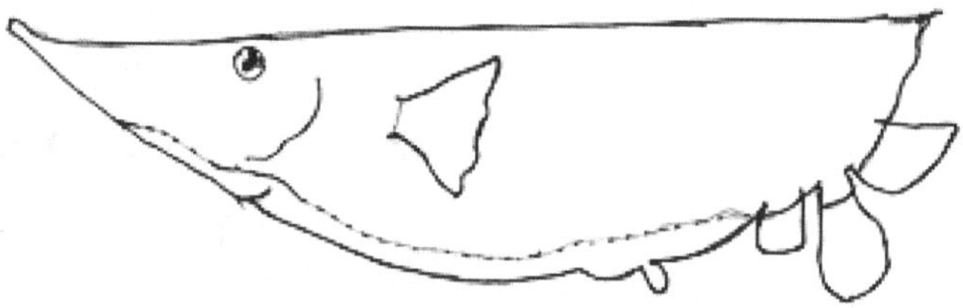

Identification Key Characters of Family: Fistularidae (Cornet fishes/ Flute mouth fishes)

- ❑ Body depressed and naked minute, line as series of suites are present
- ❑ Lateral line composed of tube-shaped ossifications that gradually take the form of long bony shields, sometimes bearing sharp spines.
- ❑ Dorsal spine absent
- ❑ CF is forked with Elongate filament.
- ❑ Lateral line present
- ❑ Mouth tube like

Layout

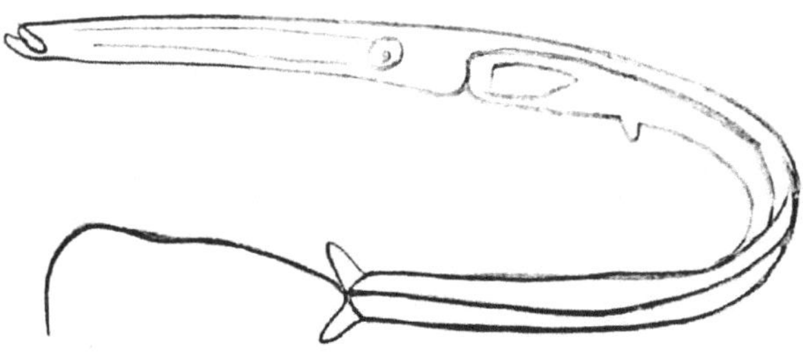

Identification Key Characters of Family: Pegasidae (Sea moth)

- ❑ Body broad and depressed
- ❑ Long rostrum is present.
- ❑ Small mouth under the rostrum
- ❑ Very large Pectoral Fins
- ❑ Caudal fin very small

Layout

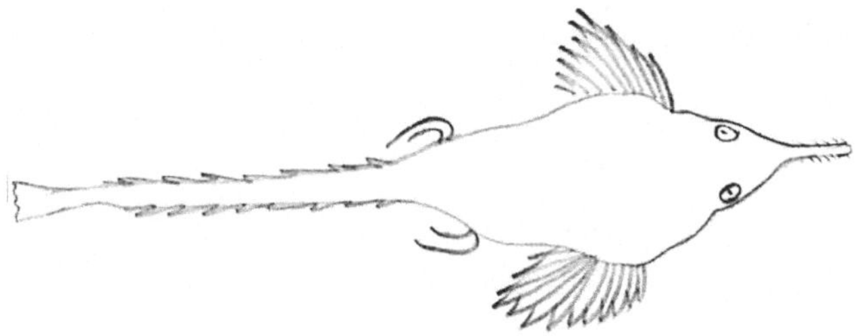

Identification Key Characters of Family: Aulostomidae (Trumpetfishes)

- ❑ Body compressed, Elongate, scaly.
- ❑ Small fleshy barbel present at tip of the lower jaw
- ❑ Small isolated dorsal spines
- ❑ Anal and second dorsal are far posterior.

Layout

INFERENCE

Chapter-28

Identification of Order – Cypriniformes (Fresh Water)

Introduction

Cypriniformes, a category of ray-finned fish, encompasses numerous families and genera, including cyprinid species like barbs, gobies, loaches, botias, and minnows. Positioned as an "order-within-an-order," Cypriniformes falls under the superorder Ostariophysi, which comprises primarily cyprinid and ostariophysin fish species.

Aim

The aim of a practical exercise focused on the identification of the order: Cypriniformes.

Systematic Classification

Kingdom: Animalia

Phylum: Chordata

Class: Actinopterygii

Order: Cypriniformes

Family: Cyprinidae, Parapsilorhynchidae, Balitoridae, Cobitidae and Psilorhynchidae.

Identification Key Characters of Family: Cyprinidae (Carps, Minnows)

- Body with scales, generally compressed abdomen rounded or cutting
- Eyes never covered with skin
- Lips usually thin, not with papillae, developed in various stages. Sometimes absent from
- one of the jaws (or) closely infesting both jaws or reflected from one or the other
- A labial fold continuous (or) interrupted present. Occasionally a few genera have a cartilaginous covering to either of the lips or both
- No subocular spines. Opercular bones well developed
- Barbels one or two pairs present (or) absent
- Last undivided dorsal fin ray osseous or articulated
- No adipose dorsal fin
- Pharyngeal teeth one to 3 rows, never more than eight teeth in any one row
- Lateral line complete (or) incomplete
- Air bladder usually large and divided into anterior and posterior chamber not surrounded by a bony capsule.

Layout

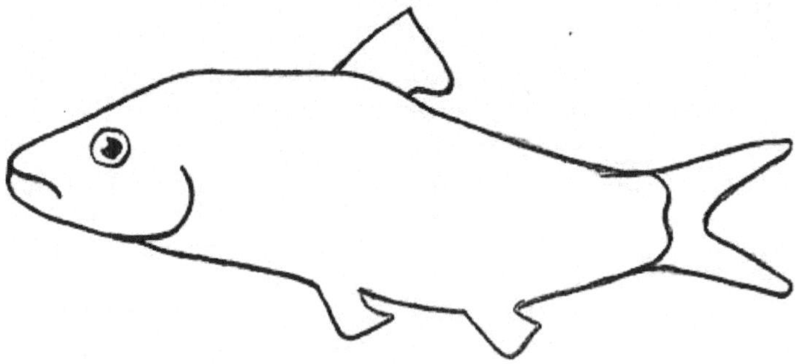

Identification Key Characters of Family: Parasilorhynchidae

- Body spindle-shaped with flat ventral surface
- Short, compressed head with pointed operculum
- Very small, inferior mouth
- Lower jaw with sharp edges and prominent lips
- Single rostral pair of barbels
- Fairly prominent lateral eyes
- Anal fin with 5 branched rays
- Dorsal fin with 8 branched rays.

Layout

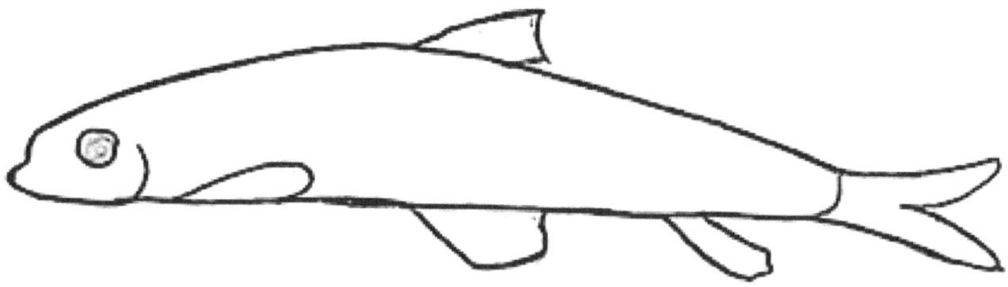

Identification Key Characters of Family: Cobitidae

- Body short, low,more or less cylindrical, posteriorly compressed but never depressed
- Body with small cycloid scale when present, usually immersed in mucus, may be present on head also
- Mouth subterminal to inferior
- Lips thick, fleshy and papillated
- 3 pairs of barbells
- An erectile spine before or below eye present
- Verticle fins with out any spine
- Pelvic fins may be absent
- Two sub families with eight genera

Layout

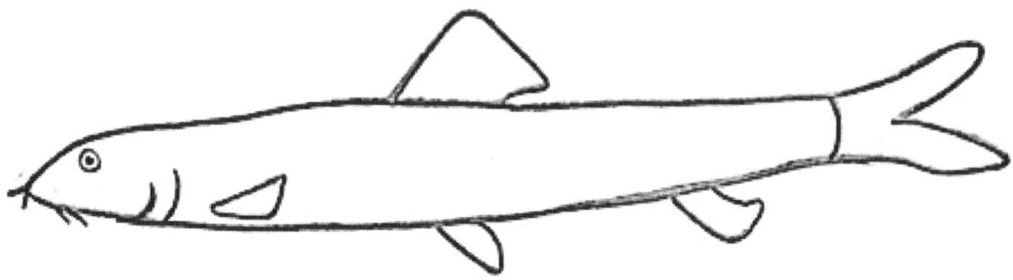

Identification Key Characters of Family: Balitoridae

- Body covered with small cycloid scales
- Depressed ventral surface, some without scales
- Gill openings restricted or moderate, united membranes
- Absence of pseudobranchiae
- Pharyngeal teeth vary, some with an adipose ridge

- ❑ Paired fins may or may not be horizontally placed
- ❑ Complete lateral line
- ❑ Elongated operculum, reduced preoperculum
- ❑ Reduced, divided air bladder in bony capsule.

Layout

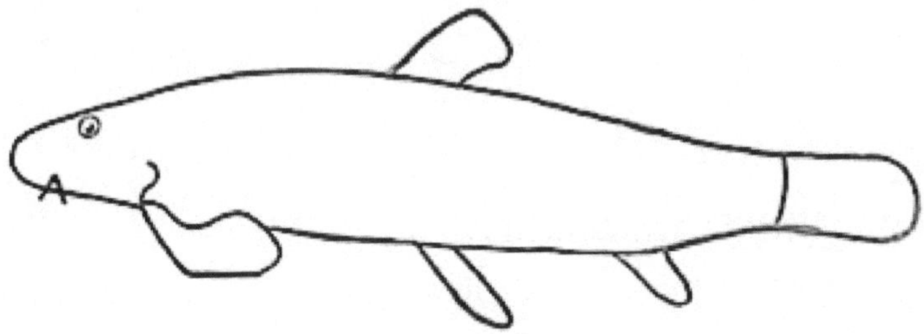

Identification Key Characters of Family: Psilorhynchidae

- ❑ **Body Shape:** Typically slender and elongated.
- ❑ **Scales:** Generally have small, cycloid scales covering the body.
- ❑ **Mouth:** Positioned ventrally, with the lower jaw usually protruding slightly.
- ❑ **Jaws:** Mouth usually contains well-developed, conical teeth.
- ❑ **Dorsal Fin:** Positioned closer to the caudal fin than the head, often with a long base.
- ❑ **Caudal Fin:** Typically deeply forked, aiding in swift swimming.
- ❑ **Coloration:** Often displaying vibrant colors or distinctive patterns, especially in males during breeding seasons.
- ❑ **Size:** Generally small in size, with most species ranging from a few centimeters to a maximum of around 15 centimeters.

Layout

INFERENCE:

Chapter-29

Identification of Order – Siluriformes (Fresh Water)

Introduction

Siluriformes, commonly known as catfish, primarily consist of freshwater species, with the exception of the mainly marine families Ariidae and Plotosidae. However, representatives of Siluriformes also inhabit brackish and freshwater environments. These fish are distributed across virtually every continent, except for Australia, showcasing their wide-ranging presence in diverse aquatic ecosystems.

Aim

The aim of a practical exercise focused on the identification of the order: Siluriformes

Systematic classification

Kingdom: Animalia

Phylum: Chordata

Class: Actinopterygii

Order: Siluriformes

Family: Ictaluridae, Bagridae, Siluridae, Pangasiidae, Sisoridae, Clariidae, Malapteruridae, Loricariidae, Heteropneustidae, Amblycipitidae, Akysidae and Schilbeidae.

Identification Key Characters of Family: Ictaluridae (American catfishes)

- North American freshwater catfishes
- Freshwater; North America (Southern Canada to Guatemala)
- Four pairs of barbels on head; skin naked.
- Dorsal and pectoral fins with a spine
- Dorsal fin usually with six soft rays.

Layout

Identification Key Characters of Family: Bagridae (Bagrid catfishes)

- Freshwater, distributed in Africa and Asia (to Japan & Borneo)
- Dorsal fin has one spine along with 6 or 7 soft rays
- Adipose fin present and highly variable in size between species.

Layout

Identification Key Characters of Family: Pangasiidae (Shark catfishes)

- Freshwater, distributed in southern Asia (Pakistan to Borneo)
- Usually two pairs of barbells, body compressed
- Adipose fin present, small and never confluent with caudal fin.

Layout

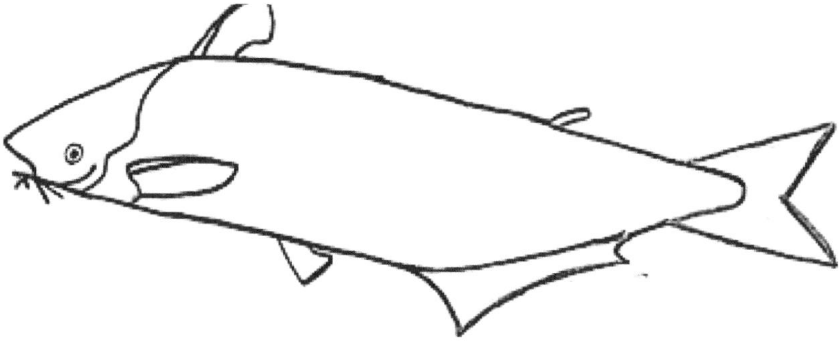

Identification Key Characters of Family: Sisoridae (Sisorid catfishes)

- Freshwater, distributed in Southern Asia
- Body usually with small unculiferous tubercles
- Thick leathery skin
- Adipose fin present
- Dorsal fin base short
- Four pairs of barbells

Layout

Identification Key Characters of Family: Clariidae (Air breathing catfishes)

- Freshwater fishes; distributed through Africa, Syria, southern and western Asia
- Dorsal fin base very long, with more than 30 rays
- Pectoral and pelvic fins absent in some species
- Four pairs of barbels present.

Layout

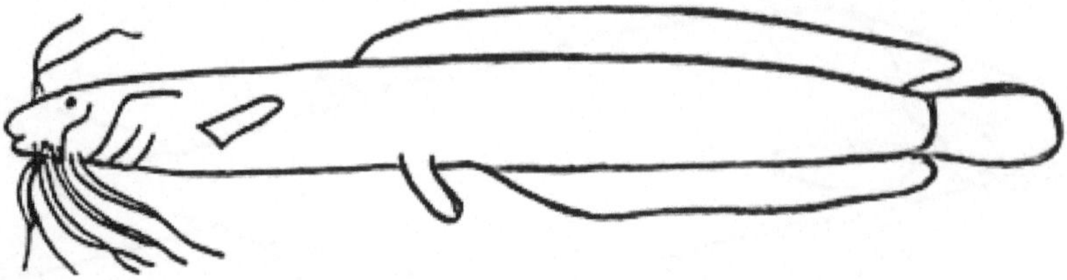

Identification Key Characters of Family: Malapteruridae Electric catfishes)

- Freshwater, distributed in Africa and Nile
- Well developed Electrogenic organ present
- Dorsal fin absent; Adipose fin far back
- Caudal fin rounded
- 3 pairs of barbells (nasal pair absent).

Layout

Identification Key Characters of Family: Loricariidae (Aromored catfishes)

- Freshwater; distributed in Panama and South America
- Body with bony plates
- Mouth ventral, with or without noticeable barbells
- Ventral lip papillose
- If adipose fin present, usually with a spine.

Layout

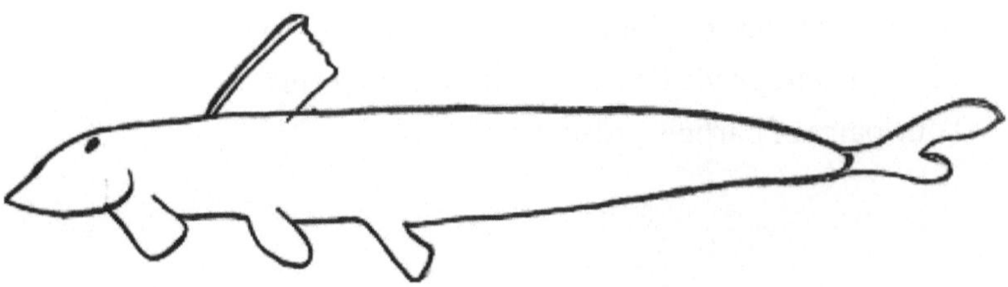

Identification Key Characters of Family: Heteropneustidae (Air sac catfish)

- Commonly known as Sting fishes, also air-breathing with tubular air sacs
- **Two known species:** *H. fossilis* and *H. longipectoralis*
- Barbels 4 pairs.

Layout

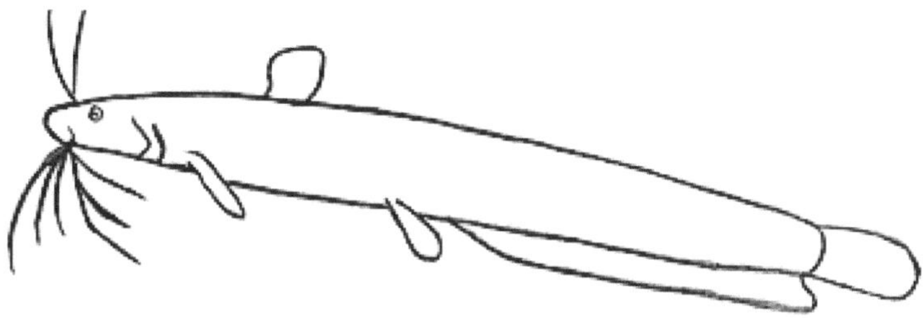

Identification Key Characters of Family: Amblycipitidae (Torrent catfishes)

- Characterized by the presence of a fold of skin above anterior to pectoral fin base
- Single genus Amblyceps with ten species
- 4 pairs of barbels present.

Layout

Identification Key Characters of Family: Akysidae (Stream catfishes)

- Pectoral fins placed horizontally with a strong osseous spine, furrowed longitudinally, serrated or smooth, covered with thick skin
- A single genus with four species

Layout

Identification Key Characters of Family: Schilbeidae(Schilbid catfishes)

- Widely separated nostrils, along the border of snout, posterior one slit-like with a flap
- Barbels elongated, one, two, or four pairs
- Anal fin very long, free from caudal fin
- 4 Pairs of barbels present.

Layout

INFERENCE:

Chapter-30

Identification of Order – Channiformes (Fresh Water)

Introduction

Channiformes, known as snakeheads, are predatory freshwater fish with streamlined bodies and scaleless skin. They possess a unique labyrinth organ that allows them to breathe air, enabling survival in oxygen-deprived waters. With sharp teeth and powerful jaws, they are adept hunters, found mainly in tropical and subtropical regions. Snakeheads are notorious for their potential invasiveness when introduced to new ecosystems, raising concerns in conservation efforts.

Aim

The aim of a practical exercise is focused on the identification of the order: Channiformes

Systematic Classification

Kingdom: Animalia

Phylum: Chordata

Class: Actinopterygii

Order: Channiformes

Family: Channidae

Identification Key Characters of Family: Channidae (Murrels)

- ❑ Body elongate, subcylindrical anteriorly.
- ❑ Body with large scales; head with plate-like scales
- ❑ Cephalic pits present.
- ❑ Teeth on jaws, vomer, and palate.
- ❑ Branchiostegals rays five.
- ❑ Pseudobranchiae absent.
- ❑ Gills four.
- ❑ Gill membranes connected beneath isthmus
- ❑ Suprabranchial organ present
- ❑ Single long spineless dorsal fin present, and a similar anal fin, both free from caudal fin
- ❑ Lateral line abruptly curved or almost interrupted
- ❑ Air-bladder present, not bifid posteriorly.

Layout

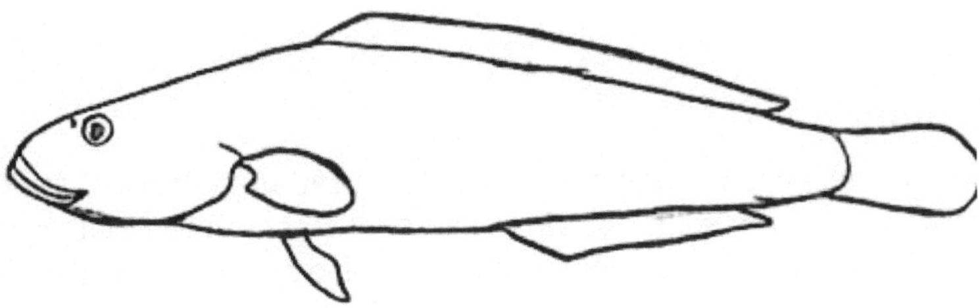

INFERENCE:

Chapter-31

Identification of Order – Cichliformes

Introduction

Cichliformes encompasses a diverse group of freshwater and marine fish, with cichlids being the most well-known family. These fish are admired for their varied colors, behaviors, and breeding strategies, making them popular in aquariums. Typically possessing a compressed body shape and large mouths with specialized teeth, cichlids exhibit complex social behaviors and adaptability to various habitats, including freshwater, brackish, and marine environments.

Aim

The aim of a practical exercise is focused on the identification of the order: Cichliformes

Systematic Classification

Kingdom: Animalia

Phylum: Chordata

Class: Actinopterygii

Order: Cichliformes

Family: Cichlidae and Pholidichthyidae

Identification Key Characters of Family: Cichlidae (Cichlids)

- ❑ Body shape quite variable, mostly moderately deep and compressed.
- ❑ A nostril on each side of head.
- ❑ Interrupted lateral line in most species.
- ❑ Scales in lateral lines may be over 100, usually 20-50.
- ❑ Dorsal fin usually with 7-25 spines and 5-30 soft rays.
- ❑ Spines in anal fin 3-15 (generally 3); soft rays 4-15 (a few with 30).
- ❑ Subocular shelf absent.

Layout

Identification Key Characters of Family: Pholidichthyidae (Convict blenny)

- ❑ A nostril to each side of head. Scaleless.
- ❑ Pelvic fins below or a little before pectoral base.
- ❑ One weak spine in pelvic fin; soft rays 2 or 3.
- ❑ Caudal fin confluent with anal and dorsal fins.
- ❑ Soft rays in dorsal fin 70-79.
- ❑ Anal soft rays 55-62.
- ❑ Pectoral fin rays 15.
- ❑ Lower pharyngeal bones coalesced into one.

Layout

INFERENCE:

Chapter-32

Identification of Order – Anabantiformes

Introduction

Anabantiformes, known as labyrinth fish, are freshwater species recognized for their labyrinth organ, enabling them to breathe air. Primarily found in tropical regions like Southeast Asia and Africa, they thrive in oxygen-deprived waters. Popular members of this order include bettas and gouramis, cherished for their vibrant colors and elaborate fins in aquariums worldwide. Anabantiformes exhibit unique reproductive strategies, such as bubble nesting.

Aim

The aim of a practical exercise is focused on the identification of the order: Anabantiformes

Systematic Classification

Kingdom: Animalia

Phylum: Chordata

Class: Actinopterygii

Order: Anabantifomes

Family: Anabantidae

Identification Key Characters of Family: Anabantidae (Climbing gouramies)

- ❑ Fixed conical teeth on jaws, prevomer, and parasphenoid.
- ❑ Relatively large mouth.
- ❑ Upper jaw slighlty protractile.
- ❑ The genus Sandelia has only cycloid scales, few gill rakers, and generally a carnivorous diet.

Layout

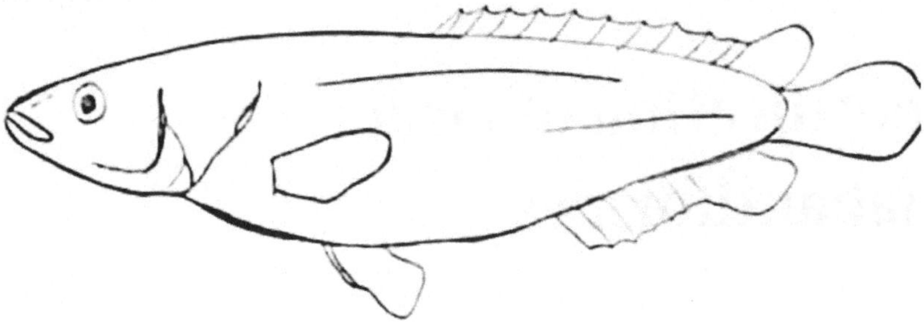

INFERENCE:

Chapter-33

General Procedure of DNA Barcoding

Introduction

Identification and classification of living organisms at the species level is crucial for biodiversity analysis and conservation. Traditional methods relying on morphological features face challenges with species discovery and differentiation. DNA barcoding, utilizing molecular techniques, addresses these challenges by providing a reliable tool for species identification and uncovering hidden diversity.

Aim

The aim of this exercise is to demonstrate the practical application of DNA barcoding for biodiversity assessment in air quality monitoring.

Materials Required

- DNA extraction kits
- PCR (Polymerase Chain Reaction) machines
- Agarose gel electrophoresis equipment
- DNA barcode primers
- PCR reagents (buffer, dNTPs, Taq polymerase)
- Microcentrifuge tubes
- Pipettes and pipette tips

- Gel documentation system
- DNA ladder
- Gel staining solution (e.g., ethidium bromide)
- Safety equipment (lab coats, gloves, goggles)
- Specimen collection tools (air samplers, swabs)
- DNA barcode reference databases
- Computer with bioinformatics software for sequence analysis

Procedure

Sample DNA Extraction

- Collect a small sample of fish from the landing area and extract tissue samples from it.
- Use a DNA extraction kit or protocol to extract DNA from the sample.
- Purify the extracted DNA using suitable methods.

PCR Amplification of DNA Barcode

- Design PCR primers targeting a specific DNA barcode region (e.g., COI gene for animals or rbcL gene for plants).
- Set up PCR reactions using the extracted DNA as a template and the designed primers.
- Optimize PCR conditions for efficient amplification.

Gel Electrophoresis

- Perform agarose gel electrophoresis to visualize the PCR products.
- Analyze the gel to confirm the presence of the amplified DNA fragments of the expected size.

DNA Sequencing

- Select a PCR product for sequencing based on the gel electrophoresis results.
- Submit the PCR product for Sanger sequencing at a sequencing facility.

Sequence Analysis

- Receive the sequencing results and perform sequence analysis using bioinformatics tools.
- Compare the obtained sequences with reference sequences in online databases such as GenBank or BOLD (Barcode of Life Database).

Species Identification

- Identify the species of the sample based on the sequence similarity with known species in the databases.

- ❑ Determine if the sample corresponds to a known species or if it represents a potentially new species.

Meta-barcoding Analysis

- ❑ Explore meta-barcoding by analysing a mixed DNA sample (e.g., soil, water) collected from a local environment.
- ❑ Use NGS (Next Generation Sequencing) technology to sequence the mixed DNA sample.
- ❑ Analyse the sequencing data to identify the different species present in the sample.

INFERENCE:

Chapter-34

General Procedure of Karyotyping

Aim

The aim of this excercisee is to know about the fundamental technique of karyotyping, which involves visualizing and analyzing the chromosomal composition of a cell sample.

Karyotyping

- The first step in the characterization of a complete chromosome set is usually known as karyotyping. It is normally followed by the measurement of each chromosome and representation of entire set of chromosomes in an internationally accepted karyotype /karyogram/idiogram.
- In fishes the chromosome classification on the basis of arm ratio proposed by Levan et al (1964) is being widely used.
- Once the karyotype has been prepared, a diagrammatic representation of the chromosome or idiogram can be prepared based on relative length of chromosome or idiogram can be presented on the basis of relative length of chromosomes.

Chromosome Classification

Centromere position	Arm ratio	Chromosome type
Median	1.00-1.07	Metacentric(m)
Sub median	1.71 - 3.00	Submetacentric(sm)
Sub terminal	3.01-7.00	Subtelocentric(st)
Terminal	Above 7.00	Acrocentric(t)

- A chromosome in the mitotic metaphase has two distinguishing features: its length and a transverse constriction that Mark's the position of the centromere. From the chromosome length and centromere position three factors are calculated: the centromeric index, arm ratio and the relative length.
- The first two factors, centromere index and arm ratio, describe about the chromosome itself and the relative length denotes about the size of chromosome in relation to other chromosomes in the set. Centromere index is defined as the length of shorter arm of the chromosome multiplied by 100 divided by the length of whole chromosomes.
- Arm ratio is defined as the length of longer arm of the chromosome divided by the length of shorter one. It is always greater than one. The relative length of a chromosome is defined as the length of whole chromosome multiplied by 100 and divided by the total length of all the chromosomes in the haploid set including the one being measured.

Protocol for Preparation of Karyotype

- Cut individual chromosomes from the photo prints.
- Group the chromosomes into four categories: Metacentric (m), submetacentric (sm) ,sub telocentric(st) & telocentric (t)
- Paste the chromosomes on ivory sheet in decreasing order of size within the group (centromeres of all chromosomes should be aligned in each row)
- Photograph the karyotype, which can be used as a base line data for detection of chromosome aberrations.
- Paste the chromosomes on Ivory sheet (Chromosomes arranged based on Centromere position)

Inference:

Chapter-35

Analysis of Protein

Introduction

Proteins are polymers of amino acids. Twenty different types of amino acids occur naturally in proteins. Proteins differ from each other according to the type, number and sequence of amino acids that make up the polypeptide backbone. As a result they have different molecular structures, nutritional attributes and physiochemical properties. Proteins are important constituents of foods for a number of different reasons. They are a major source of *energy*, as well as containing essential amino-acids, such as lysine, tryptophan, methionine, leucine, isoleucine and valine, which are essential to human health, but which the body cannot synthesize. Proteins are also the major structural components of many natural foods, often determining their overall texture, *e.g.,* tenderness of meat or fish products. Isolated proteins are often used in foods as ingredients because of their unique functional properties, *i.e.,* their ability to provide desirable appearance, texture or stability. Typically, proteins are used as gelling agents, emulsifiers, foaming agents and thickeners. Many food proteins are enzymes which are capable of enhancing the rate of certain biochemical reactions. These reactions can have either a favorable or detrimental effect on the overall properties of foods. Food analysts are interested in knowing the total concentration, type, molecular structure and functional properties of the proteins in foods.

Determination of Overall Protein Concentration

Kjeldahl Method

The Kjeldahl method was developed in 1883 by a brewer called Johann Kjeldahl. A food is digested with a strong acid so that it releases nitrogen which can be determined by a suitable titration technique. The amount of protein present is then calculated from the nitrogen concentration of the food. The same basic approach is still used today, although a number of improvements have been made to speed up the process and to obtain more accurate measurements. It is usually considered to be *the* standard method of determining protein concentration. Because the Kjeldahl method does not measure the protein content directly a *conversion factor (F)* is needed to convert the measured nitrogen concentration to a protein concentration. A conversion factor of 6.25 (equivalent to 0.16 g nitrogen per gram of protein) is used for many applications, however, this is only an average value, and each protein has a different conversion factor depending on its amino-acid composition. The Kjeldahl method can conveniently be divided into three steps: digestion, neutralization and titration.

Principles

Digestion

The food sample to be analyzed is weighed into a *digestion flask* and then digested by heating it in the presence of sulfuric acid (an oxidizing agent which digests the food), anhydrous sodium sulfate (to speed up the reaction by raising the boiling point) and a catalyst, such as copper, selenium, titanium, or mercury (to speed up the reaction). Digestion converts any nitrogen in the food (other than that which is in the form of nitrates or nitrites) into ammonia, and other organic matter to $C0_2$ and H_20. Ammonia gas is not liberated in an acid solution because the ammonia is in the form of the ammonium ion (NH_4^+) which binds to the sulfate ion (SO_4^{2-}) and thus remains in solution:

$$N(food) \rightarrow (NH_4)_2SO_4 \ (1)$$

Neutralization

After the digestion has been completed the digestion flask is connected to a *recieving flask* by a tube. The solution in the digestion flask is then made alkaline by addition of sodium hydroxide, which converts the ammonium sulfate into ammonia gas:

$$(NH_4)_2SO_4 + 2\ NaOH \rightarrow 2NH_3 + 2H_2O + Na_2SO_4 \ (2)$$

The ammonia gas that is formed is liberated from the solution and moves out of the digestion flask and into the receiving flask - which contains an excess of boric acid. The low pH of the solution in the receiving flask converts the ammonia gas into the ammonium ion, and simultaneously converts the boric acid to the borate ion:

$$NH_3 + H_3BO_3 \text{ (boric acid)} \rightarrow NH_4^+ + H_2BO_3^- \text{ (borate ion)} \ (3)$$

Titration

The nitrogen content is then estimated by titration of the ammonium borate formed with standard sulfuric or hydrochloric acid, using a suitable indicator to determine the end-point of the reaction.

$$H_2BO_3^- + H^+ \rightarrow H_3BO_3 \quad (4)$$

The concentration of hydrogen ions (in moles) required to reach the end-point is equivalent to the concentration of nitrogen that was in the original food (Equation 3). The following equation can be used to determine the nitrogen concentration of a sample that weighs *m* grams using a *x*M HCl acid solution for the titration:

$$\% N = \frac{x \text{ moles}}{1000 \text{ cm}^3} \times \frac{(v_s - v_b) \text{ cm}^3}{m \text{ g}} \times \frac{14 \text{ g}}{\text{moles}} \times 100 \quad \textbf{(5)}$$

Where *vs* and *vb* are the titration volumes of the sample and blank, and 14g is the molecular weight of nitrogen N. A blank sample is usually ran at the same time as the material being analyzed to take into account any residual nitrogen which may be in the reagents used to carry out the analysis. Once the nitrogen content has been determined it is converted to a protein content using the appropriate conversion factor: %Protein = F %N.

Advantages and Disadvantages

Advantages. The Kjeldahl method is widely used internationally and is still the standard method for comparison against all other methods. Its universality, high precision and good reproducibility have made it the major method for the estimation of protein in foods.

Disadvantages. It does not give a measure of the true protein, since all nitrogen in foods is not in the form of protein. Different proteins need different correction factors because they have different amino acid sequences. The use of concentrated sulfuric acid at high temperatures poses a considerable hazard, as does the use of some of the possible catalysts The technique is time consuming to carry-out.

Enhanced Dumas Method

Recently, an automated instrumental technique has been developed which is capable of rapidly measuring the protein concentration of food samples. This technique is based on a method first described by a scientist called Dumas over a century and a half ago. It is beginning to compete with the Kjeldahl method as the standard method of analysis for proteins for some foodstuffs due to its rapidness.

General Principles

A sample of known mass is combusted in a high temperature (about 900 °C) chamber in the presence of oxygen. This leads to the release of CO_2, H_2O and N_2. The CO_2 and H_2O are removed by passing the gasses over special columns that absorb them. The nitrogen content is then measured by passing the remaining gasses through a column that has a thermal conductivity detector at the end. The column helps separate the nitrogen from any residual CO_2 and H_2O that may have remained in the gas stream. The instrument is calibrated by analyzing a material that is pure and has a known nitrogen concentration, such as EDTA (= 9.59%N). Thus the signal from the thermal conductivity detector can be converted into a nitrogen content. As with the Kjeldahl method it is necessary to convert the concentration of nitrogen in a sample to the protein content, using suitable conversion factors which depend on the precise amino acid sequence of the protein.

Advantages and Disadvantages

Advantages: It is much faster than the Kjeldahl method (under 4 minutes per measurement, compared to 1-2 hours for Kjeldahl). It doesn't need toxic chemicals or catalysts. Many samples can be measured automatically. It is easy to use.

Disadvantages: High initial cost. It does not give a measure of the true protein, since all nitrogen in foods is not in the form of protein. Different proteins need different correction factors because they have different amino acid sequences. The small sample size makes it difficult to obtain a representative sample.

Methods using UV-visible Spectroscopy

A number of methods have been devised to measure protein concentration, which are based on UV-visible spectroscopy. These methods use either the natural ability of proteins to absorb (or scatter) light in the UV-visible region of the electromagnetic spectrum, or they chemically or physically modify proteins to make them absorb (or scatter) light in this region. The basic principle behind each of these tests is similar. First of all a calibration curve of absorbance (or turbidity) versus protein concentration is prepared using a series of protein solutions of known concentration. The absorbance (or turbidity) of the solution being analyzed is then measured at the same wavelength, and its protein concentration determined from the calibration curve. The main difference between the tests are the chemical groups which are responsible for the absorption or scattering of radiation, *e.g.*, peptide bonds, aromatic side-groups, basic groups and aggregated proteins.

A number of the most commonly used UV-visible methods for determining the protein content of foods are highlighted below:

Principles

Direct measurement at 280nm

Tryptophan and tyrosine absorb ultraviolet light strongly at 280 nm. The tryptophan and tyrosine content of many proteins remains fairly constant, and so the absorbance of protein solutions at 280nm can be used to determine their concentration. The advantages of this method are that the procedure is simple to carry out, it is nondestructive, and no special reagents are required. The major disadvantage is that nucleic acids also absorb strongly at 280 nm and could therefore interfere with the measurement of the protein if they are present in sufficient concentrations. Even so, methods have been developed to overcome this problem, *e.g.*, by measuring the absorbance at two different wavelengths.

Biuret Method

A violet-purplish color is produced when cupric ions (Cu^{2+}) interact with *peptide bonds* under alkaline conditions. The biuret reagent, which contains all the chemicals required to carry out the analysis, can be purchased commercially. It is mixed with a protein solution and then allowed to stand for 15-30 minutes before the absorbance is read at 540 nm. The major advantage of this technique is that there is no interference from materials that adsorb at lower wavelengths, and the technique is less sensitive to protein type because it utilizes absorption involving peptide bonds that are common to all proteins, rather than specific side groups. However, it has a relatively low sensitivity compared to other UV-visible methods.

Lowry Method

The Lowry method combines the biuret reagent with another reagent (the Folin-Ciocalteau phenol reagent) which reacts with *tyrosine and tryptophan* residues in proteins. This gives a bluish color which can be read somewhere between 500 - 750 nm depending on the sensitivity required. There is a small peak around 500 nm that can be used to determine high protein concentrations and a large peak around 750 nm that can be used to determine low protein concentrations. This method is more sensitive to low concentrations of proteins than the biuret method.

Dye binding Methods

A known excess of a negatively charged (anionic) dye is added to a protein solution whose pH is adjusted so that the proteins are positively charged (*i.e.* < the isoelectric point). The proteins form an insoluble complex with the dye because of the electrostatic attraction between the molecules, but the unbound dye remains soluble. The anionic dye binds to cationic groups of the basic amino acid residues (histidine, arganine and lysine) and to free amino terminal groups. The amount of unbound dye remaining in solution after the insoluble protein-dye complex has been removed (*e.g.*, by centrifugation) is determined by measuring its absorbance. The amount of protein present in the original solution is proportional to the amount of dye that bound to it: $dye_{bound} = dye_{initial} - dye_{free}$.

Turbimetric Method

Protein molecules which are normally soluble in solution can be made to precipitate by the addition of certain chemicals, *e.g.*, trichloroacetic acid. Protein precipitation causes the solution to become turbid. Thus the concentration of protein can be determined by measuring the degree of turbidity.

Advantages and Disadvantages

Advantages: UV-visible techniques are fairly rapid and simple to carry out, and are sensitive to low concentrations of proteins.

Disadvantages: For most UV-visible techniques it is necessary to use dilute and transparent solutions, which contain no contaminating substances which absorb or scatter light at the same wavelength as the protein being analyzed. The need for transparent solutions means that most foods must undergo significant amounts of sample preparation before they can be analyzed, *e.g.*, homogenization, solvent extraction, centrifugation, filtration, which can be time consuming and laborious. In addition, it is sometimes difficult to quantitatively extract proteins from certain types of foods, especially after they have been processed so that the proteins become aggregated or covalently bound with other substances. In addition the absorbance depends on the type of protein analyzed (different proteins have different amino acid sequences).

Other Instrumental Techniques

There are a wide variety of different instrumental methods available for determining the total protein content of food materials. These can be divided into three different categories according to their physicochemical principles: (i) measurement of bulk physical properties, (ii) measurement of adsorption of radiation, and (iii) measurement of scattering of radiation. Each instrumental methods has its own advantages and disadvantages, and range of foods to which it can be applied.

Principles

Measurement of Bulk Physical Properties

- *Density:* The density of a protein is greater than that of most other food components, and so there is an increase in density of a food as its protein content increases. Thus the protein content of foods can be determined by measuring their density.
- *Refractive index:* The refractive index of an aqueous solution increases as the protein concentration increases and therefore RI measurements can be used to determine the protein content.

Measurement of Adsorption of Radiation

- *UV-visible:* The concentration of proteins can be determined by measuring the absorbance of ultraviolet-visible radiation (see above).
- *Infrared:* Infrared techniques can be used to determine the concentration of proteins in food samples. Proteins absorb IR naturally due to characteristic vibrations (stretching and bending) of certain chemical groups along the polypeptide backbone. Measurements of the absorbance of radiation at certain wavelengths can thus be used to quantify the concentration of protein in the sample. IR is particularly useful for rapid on-line analysis of protein content. It also requires little sample preparation and is nondestructive. Its major disadvantages are its high initial cost and the need for extensive calibration.

- *Nuclear Magnetic Resonance*: NMR spectroscopy can be used to determine the total protein concentration of foods. The protein content is determined by measuring the area under a peak in an NMR chemical shift spectra that corresponds to the protein fraction.

Measurement of Scattering of Radiation

- *Light scattering:* The concentration of protein aggregates in aqueous solution can be determined using light scattering techniques because the turbidity of a solution is directly proportional to the concentration of aggregates present.
- *Ultrasonic scattering:* The concentration of protein aggregates can also be determined using ultrasonic scattering techniques because the ultrasonic velocity and absorption of ultrasound are related to the concentration of protein aggregates present.

Advantages and Disadvantages

A number of these instrumental methods have major advantages over the other techniques mentioned above because they are nondestructive, require little or no sample preparation, and measurements are rapid and precise. A major disadvantage of the techniques which rely on measurements of the bulk physical properties of foods are that a calibration curve must be prepared between the physical property of interest and the total protein content, and this may depend on the type of protein present and the food matrix it is contained within. In addition, the techniques based on measurements of bulk physicochemical properties can only be used to analyze foods with relatively simple compositions. In a food that contains many different components whose concentration may vary, it is difficult to disentangle the contribution that the protein makes to the overall measurement from that of the other components.

Comparison of Methods

As food scientists we may often be in a position where we have to choose a particular technique for measuring the protein concentration of a food. How do we decide which technique is the most appropriate for our particular application ? The first thing to determine is what is the information going to be used for. If the analysis is to be carried out for official purposes, *e.g.*, legal or labeling requirements, then it is important to use an officially recognized method. The Kjeldahl method, and increasingly the Dumas method, have been officially approved for a wide range of food applications. In contrast, only a small number of applications of UV-visible spectroscopy have been officially recognized.

For quality control purposes, it is often more useful to have rapid and simple measurements of protein content and therefore IR techniques are most suitable. For fundamental studies in the laboratory, where pure proteins are often analyzed, UV-visible spectroscopic techniques are often preferred because they give rapid and reliable measurements, and are sensitive to low concentrations of protein.

Other factors which may have to be considered are the amount of sample preparation required, their sensitivity and their speed. The Kjeldahl, Dumas and IR methods require very little sample preparation. After a representative sample of the food has been selected it can usually be tested directly. On the other hand, the various UV-visible methods require extensive sample preparation prior to analysis. The protein must be extracted from the food into a dilute transparent solution, which usually involves time consuming homogenization, solvent extraction, filtration and centrifugation procedures. In addition, it may be difficult to completely isolate some proteins from foods because they are strongly bound to other components. The various techniques also have different sensitivities, *i.e.*, the lowest concentration of protein which they can detect. The UV-visible methods are the most sensitive, being able to detect protein concentrations as low as 0.001 wt%. The sensitivity of the Dumas, Kjeldahl and IR methods is somewhere around 0.1 wt%. The time required per analysis, and the number of samples which can be run simultaneously, are also important factors to consider when deciding which analytical technique to use. IR techniques are capable of rapid analysis (< 1 minute) of protein concentration once they have been calibrated. The modern instrumental Dumas method is fully automated and can measure the protein concentration of a sample in less than 5 minutes, compared to the Kjeldahl method which takes between 30 minutes and 2 hours to carry out. The various UV-visible methods range between a couple of minutes to an hour (depending on the type of dye that is used and how long it takes to react), although it does have the advantage that many samples can be run simultaneously. Nevertheless, it is usually necessary to carry out extensive sample preparation prior to analysis in order to get a transparent solution. Other factors which may be important when selecting an appropriate technique are: the equipment available, ease of operation, the desired accuracy, and whether or not the technique is nondestructive.

Protein Separation and Characterization

In above techniques used to determine the total *concentration* of protein in a food were discussed. Food analysts are also often interested in the *type* of proteins present in a food because each protein has unique nutritional and physicochemical properties. Protein type is usually determined by separating and isolating the individual proteins from a complex mixture of proteins, so that they can be subsequently identified and characterized. Proteins are separated on the basis of differences in their physicochemical properties, such as size, charge, adsorption characteristics, solubility and heat-stability. The choice of an appropriate separation technique depends on a number of factors, including the reasons for carrying out the analysis, the amount of sample available, the desired purity, the equipment available, the type of proteins present and the cost. Large-scale methods are available for crude isolations of large quantities of proteins, whereas small-scale methods are available for proteins that are expensive or only available in small quantities. One of the factors that must be considered during the separation procedure is the possibility that the native three dimensional structure of the protein molecules may be altered.

A prior knowledge of the effects of environmental conditions on protein structure and interactions is extremely useful when selecting the most appropriate separation technique. Firstly, because it helps determine the most suitable conditions to use to isolate a particular protein from a mixture of proteins (*e.g.*, pH, ionic strength, solvent, temperature *etc.*), and secondly, because it may be important to choose conditions which will not adversely affect the molecular structure of the proteins.

Methods Based on Different Solubility Characteristics

Proteins can be separated by exploiting differences in their solubility in aqueous solutions. The solubility of a protein molecule is determined by its amino acid sequence because this determines its size, shape, hydrophobicity and electrical charge. Proteins can be selectively precipitated or solubilized by altering the pH, ionic strength, dielectric constant or temperature of a solution. These separation techniques are the most simple to use when large quantities of sample are involved, because they are relatively quick, inexpensive and are not particularly influenced by other food components. They are often used as the first step in any separation procedure because the majority of the contaminating materials can be easily removed.

Salting Out

Proteins are precipitated from aqueous solutions when the salt concentration exceeds a critical level, which is known as *salting-out*, because all the water is "bound" to the salts, and is therefore not available to hydrate the proteins. Ammonium sulfate [$(NH_4)_2SO_4$] is commonly used because it has a high water-solubility, although other neutral salts may also be used, *e.g.*, NaCl or KCl. Generally a two-step procedure is used to maximize the separation efficiency. In the first step, the salt is added at a concentration just below that necessary to precipitate out the protein of interest. The solution is then centrifuged to remove any proteins that are less soluble than the protein of interest. The salt concentration is then increased to a point just above that required to cause precipitation of the protein. This precipitates out the protein of interest (which can be separated by centrifugation), but leaves more soluble proteins in solution. The main problem with this method is that large concentrations of salt contaminate the solution, which must be removed before the protein can be resolubilzed, *e.g.*, by dialysis or ultrafiltration.

Isoelectric Precipitation

The *isoelectric point* (pI) of a protein is the pH where the net charge on the protein is zero. Proteins tend to aggregate and precipitate at their pI because there is no electrostatic repulsion keeping them apart. Proteins have different isoelectric points because of their different amino acid sequences (*i.e.*, relative numbers of anionic and cationic groups), and thus they can be separated by adjusting the pH of a solution. When the pH is adjusted to the pI of a particular protein it precipitates leaving the other proteins in solution.

Solvent Fractionation

The solubility of a protein depends on the *dielectric constant* of the solution that surrounds it because this alters the magnitude of the electrostatic interactions between charged groups. As the dielectric constant of a solution decreases the magnitude of the electrostatic interactions between charged species increases. This tends to decrease the solubility of proteins in solution because they are less ionized, and therefore the electrostatic repulsion between them is not sufficient to prevent them from aggregating. The dielectric constant of aqueous solutions can be lowered by adding water-soluble organic solvents, such as ethanol or acetone. The amount of organic solvent required to cause precipitation depends on the protein and therefore proteins can be separated on this basis. The optimum quantity of organic solvent required to precipitate a protein varies from about 5 to 60%. Solvent fractionation is usually performed at 0°C or below to prevent protein denaturation caused by temperature increases that occur when organic solvents are mixed with water.

Denaturation of Contaminating Proteins

Many proteins are denatured and precipitate from solution when heated above a certain temperature or by adjusting a solution to highly acid or basic pHs. Proteins that are stable at high temperature or at extremes of pH are most easily separated by this technique because contaminating proteins can be precipitated while the protein of interest remains in solution.

Separation due to Different Adsorption Characteristics

Adsorption chromatography involves the separation of compounds by selective adsorption-desorption at a solid matrix that is contained within a column through which the mixture passes. Separation is based on the different affinities of different proteins for the solid matrix. *Affinity* and *ion-exchange* chromatography are the two major types of adsorption chromatography commonly used for the separation of proteins. Separation can be carried out using either an open column or high-pressure liquid chromatography.

Ion Exchange Chromatography

Ion exchange chromatography relies on the reversible adsorption-desorption of ions in solution to a charged solid matrix or polymer network. This technique is the most commonly used chromatographic technique for protein separation. A positively charged matrix is called an *anion-exchanger* because it binds negatively charged ions (anions). A negatively charged matrix is called a cat*ion-exchanger* because it binds positively charged ions (cations). The buffer conditions (pH and ionic strength) are adjusted to favor maximum binding of the protein of interest to the ion-exchange column. Contaminating proteins bind less strongly and therefore pass more rapidly through the column. The protein of interest is then eluted using another buffer solution which favors its desorption from the column (*e.g.*, different pH or ionic strength).

Affinity Chromatography

Affinity chromatography uses a stationary phase that consists of a *ligand* covalently bound to a solid support. The ligand is a molecule that has a highly specific and unique reversible affinity for a particular protein. The sample to be analyzed is passed through the column and the protein of interest binds to the ligand, whereas the contaminating proteins pass directly through. The protein of interest is then eluted using a buffer solution which favors its desorption from the column. This technique is the most efficient means of separating an individual protein from a mixture of proteins, but it is the most expensive, because of the need to have columns with specific ligands bound to them.

Both ion-exchange and affinity chromatography are commonly used to separate proteins and amino-acids in the laboratory. They are used less commonly for commercial separations because they are not suitable for rapidly separating large volumes and are relatively expensive.

Separation Due to Size Differences

Proteins can also be separated according to their size. Typically, the molecular weights of proteins vary from about 10,000 to 1,000,000 daltons. In practice, separation depends on the *Stokes radius* of a protein, rather than directly on its molecular weight. The Stokes radius is the average radius that a protein has in solution, and depends on its three dimensional molecular structure. For proteins with the same molecular weight the Stokes radius increases in the following order: compact globular protein < flexible random-coil < rod-like protein.

Dialysis

Dialysis is used to separate molecules in solution by use of semipermeable membranes that permit the passage of molecules smaller than a certain size through, but prevent the passing of larger molecules. A protein solution is placed in dialysis tubing which is sealed and placed into a large volume of water or buffer which is slowly stirred. Low molecular weight solutes flow through the bag, but the large molecular weight protein molecules remain in the bag. Dialysis is a relatively slow method, taking up to 12 hours to be completed. It is therefore most frequently used in the laboratory. Dialysis is often used to remove salt from protein solutions after they have been separated by salting-out, and to change buffers.

Ultrafiltration

A solution of protein is placed in a cell containing a *semipermeable membrane,* and pressure is applied. Smaller molecules pass through the membrane, whereas the larger molecules remain in the solution. The separation principle of this technique is therefore similar to dialysis, but because pressure is applied separation is much quicker. Semipermeable membranes with cutoff points between about 500 to 300,000 are available. That portion of the solution which is retained by the cell (large molecules) is called the *retentate,* whilst that part which passes through the membrane (small molecules) forms part of the *ultrafiltrate.* Ultrafiltration can be used to concentrate a protein solution, remove salts, exchange buffers or fractionate proteins on the basis of their size. Ultrafiltration units are used in the laboratory and on a commercial scale.

Size Exclusion Chromatography

This technique, sometimes known as gel filtration, also separates proteins according to their size. A protein solution is poured into a column which is packed with porous beads made of a cross-linked polymeric material (such as dextran or agarose). Molecules larger than the pores in the beads are excluded, and move quickly through the column, whereas the movement of molecules which enter the pores is retarded. Thus molecules are eluted off the column in order of decreasing size. Beads of different average pore size are available for separating proteins of different molecular weights. Manufacturers of these beads provide information about the molecular weight range that they are most suitable for separating. Molecular weights of unknown proteins can be determined by comparing their elution volumes Vo, with those determined using proteins of known molecular weight: a plot of elution volume versus log(*molecular weight*) should give a straight line. One problem with this method is that the molecular weight is not directly related to the Stokes radius for different shaped proteins.

Separation by Electrophoresis

Electrophoresis relies on differences in the migration of charged molecules in a solution when an electrical field is applied across it. It can be used to separate proteins on the basis of their size, shape or charge.

Non-denaturing Electrophoresis

In non-denaturing electrophoresis, a buffered solution of native proteins is poured onto a porous gel (usually polyacrylamide, starch or agarose) and a voltage is applied across the gel. The proteins move through the gel in a direction that depends on the sign of their charge, and at a rate that depends on the magnitude of the charge, and the friction to their movement:

$$\text{mobility} = \frac{\text{applied voltage} \times \text{molecular charge}}{\text{molecular friction}}$$

Proteins may be positively or negatively charged in solution depending on their isoelectic points (pI) and the pH of the solution. A protein is negatively charged if the pH is above the pI, and positively charged if the pH is below the pI. The magnitude of the charge and applied voltage will determine how far proteins migrate in a certain time. The higher the voltage or the greater the charge on the protein the further it will move. The friction of a molecule is a measure of its resistance to movement through the gel and is largely determined by the relationship between the effective size of the molecule, and the size of the pores in the gel. The smaller the size of the molecule, or the larger the size of the pores in the gel, the lower the resistance and therefore the faster a molecule moves through the gel. Gels with different porosity's can be purchased from chemical suppliers, or made up in the laboratory. Smaller pores sizes are obtained by using a higher concentration of cross-linking reagent to form the gel. Gels may be contained between two parallel plates, or in cylindrical tubes. In non-denaturing electrophoresis the native proteins are separated based on a combination of their charge, size and shape.

Denaturing Electrophoresis

In *denaturing* electrophoresis proteins are separated primarily on their molecular weight. Proteins are denatured prior to analysis by mixing them with mercaptoethanol, which breaks down disulfide bonds, and *sodium dodecyl sulfate* (SDS*)*, which is an anionic surfactant that hydrophobically binds to protein molecules and causes them to unfold because of the repulsion between negatively charged surfactant head-groups. Each protein molecule binds approximately the same amount of SDS per unit length. Hence, the charge per unit length and the molecular conformation is approximately similar for all proteins. As proteins travel through a gel network they are primarily separated on the basis of their molecular weight because their movement depends on the size of the protein molecule relative to the size of the pores in the gel: smaller proteins moving more rapidly through the matrix than larger molecules. This type of electrophoresis is commonly called *sodium dodecyl sulfate -polyacrylamide gel electrophoresis,* or *SDS-PAGE.*

To determine how far proteins have moved a *tracking dye* is added to the protein solution, *e.g.,* bromophenol blue. This dye is a small charged molecule that migrates ahead of the proteins. After the electrophoresis is completed the proteins are made visible by treating the gel with a protein dye such as Coomassie Brilliant Blue or silver stain. The relative mobility of each protein band is calculated:

$$R_m = \frac{\text{distance protein moves}}{\text{distance dye moves}}$$

Electrophoresis is often used to determine the protein composition of food products. The protein is extracted from the food into solution, which is then separated using electrophoresis. SDS-PAGE is used to determine the molecular weight of a protein by measuring *R*m, and then comparing it with a calibration curve produced using proteins of known molecular weight: a plot of log (molecular weight) against relative mobility is usually linear. Denaturing electrophoresis is more useful for determining molecular weights than non-denaturing electrophoresis, because the friction to movement does not depend on the shape or original charge of the protein molecules.

Isoelectric Focusing Electrophoresis

This technique is a modification of electrophoresis, in which proteins are separated by *charge* on a gel matrix which has a pH gradient across it. Proteins migrate to the location where the pH equals their isoelectric point and then stop moving because they are no longer charged. This method has one of the highest resolutions of all techniques used to separate proteins. Gels are available that cover a narrow pH range (2-3 units) or a broad pH range (3-10 units) and one should therefore select a gel which is most suitable for the proteins being separated.

Two-Dimensional Electrophoresis

Isoelectric focusing and SDS-PAGE can be used together to improve resolution of complex protein mixtures. Proteins are separated in one direction on the basis of *charge* using isoelectric focusing, and then in a perpendicular direction on the basis of *size* using SDS-PAGE.

Amino Acid Analysis

Amino acid analysis is used to determine the amino acid composition of proteins. A protein sample is first hydrolyzed (*e.g.* using a strong acid) to release the amino acids, which are then separated using chromatography, *e.g.,* ion exchange, affinity or absorption chromatography.

INFERENCE:

Protein analysis plays a significant role in fish identification by enabling the differentiation and classification of species based on their unique protein profiles. Techniques such as mass spectrometry and electrophoresis are used to analyze the protein composition of fish tissues, providing a biochemical fingerprint that can be compared across different specimens. This approach is particularly useful in cases where morphological characteristics are insufficient or ambiguous for species identification. Protein analysis not only aids in accurate species identification but also contributes to understanding the phylogenetic relationships among fish species, which is essential for biodiversity studies, conservation efforts, and the management of fishery resources.]

Reference

1. Cawthorn, D.-M., Steinman, H.A., & Hoffman, L.C. (2012). A high incidence of species substitution and mislabeling detected in meat products sold in South Africa. *Food Control*, 32(2), 440-449.

Chapter-36

Fish Species Identification Using PCR-RFLP Analysis

Introduction

Fish species identification is crucial in various industries, including food, aquaculture, and conservation. PCR-RFLP (polymerase chain reaction-restriction fragment length polymorphism) analysis is a reliable method for identifying fish species. This technique involves amplifying specific regions of the mitochondrial cytochrome b gene using PCR and then digesting the amplified DNA with restriction enzymes to generate unique profiles for each species.

Aim

The aim is to identify fish species using PCR-RFLP analysis, a reliable method that involves amplifying specific regions of the mitochondrial cytochrome b gene using PCR and then digesting the amplified DNA with restriction enzymes to generate unique profiles for each species.

Target Gene and Primers

The mitochondrial cytochrome b gene is commonly used as the target sequence for PCR-RFLP analysis due to its high conservation across fish species. Primers are designed to amplify specific regions of the cytochrome b gene, allowing for species-specific identification.

PCR and Restriction Enzymes

PCR is used to amplify the target sequence, followed by digestion with restriction enzymes to generate specific fragments. Different restriction enzymes are used to create unique profiles for each species, allowing for identification.

Lab-on-a-Chip Capillary Electrophoresis

Lab-on-a-chip capillary electrophoresis is used for end-point analysis, enabling accurate sizing of DNA fragments and identification of fish species at low concentrations.

Database and Reference Materials

A database of PCR-RFLP profiles is created and used for comparison with unknown samples, eliminating the need for reference materials. This approach ensures accurate identification of fish species without the need for morphological characteristics or protein profiling.

Applications and Interlaboratory Studies

PCR-RFLP analysis has been successfully applied to identify various fish species, including tuna, cod, anglerfish, and pufferfish. Interlaboratory studies have demonstrated high accuracy and consistency in identifying fish species using PCR-RFLP analysis.

Advantages and Limitations

PCR-RFLP analysis offers several advantages, including simplicity, low cost, and high accuracy, making it a reliable method for fish species identification. However, it may not be suitable for heavily processed samples where DNA is degraded or denatured.

Materials Required

DNA Samples: Fin tissue or mucus swabs from fish samples. Blood samples can also be used, but fin tissue is generally preferred.

PCR Reagents: DNA polymerase (e.g., Taq polymerase), dNTPs (dATP, dCTP, dGTP, dTTP), primers specific to the target gene (cytochrome b gene in this case), PCR buffer (e.g., Tris-HCl, KCl, MgCl2).

Restriction Enzymes: DdeI, NlaIII, and HaeIII are commonly used for PCR-RFLP analysis. Other enzymes like EcoRI, BamHI, and HindIII can also be used depending on the specific requirements.

Lab-on-a-Chip Capillary Electrophoresis: The lab-on-a-chip capillary electrophoresis system is used for end-point analysis to accurately size DNA fragments.

Software and Database: Software for data analysis and comparison to a database of PCR-RFLP profiles. A database of PCR-RFLP profiles for different fish species is necessary for identification.

Equipment: PCR thermocycler, agarose gel for DNA visualization, UV transillluminator for DNA visualization, pipettes and microcentrifuge tubes for handling DNA samples.

Procedure

PCR Amplification:

- ❑ Prepare the PCR reaction mixture with the target gene primers and PCR reagents.
- ❑ Perform PCR amplification using a thermocycler.
- ❑ Verify the amplification by agarose gel electrophoresis.

Restriction Enzyme Digestion:

- ❑ Prepare the restriction enzyme digestion reaction mixture with the amplified DNA and restriction enzymes.
- ❑ Perform the digestion reaction.
- ❑ Verify the digestion by agarose gel electrophoresis.

Lab-on-a-Chip Capillary Electrophoresis:

- ❑ Prepare the lab-on-a-chip capillary electrophoresis system with the digested DNA fragments.
- ❑ Perform the capillary electrophoresis analysis.
- ❑ Analyze the results using software for data analysis and comparison to a database of PCR-RFLP profiles.

Database Comparison:

- ❑ Compare the PCR-RFLP profiles generated from the unknown sample to the database of PCR-RFLP profiles for different fish species.
- ❑ Identify the species based on the best match.

Conclusion

PCR-RFLP analysis is a reliable method for identifying fish species. The technique involves amplifying specific regions of the mitochondrial cytochrome b gene using PCR and then digesting the amplified DNA with restriction enzymes to generate unique profiles for each species. The use of lab-on-a-chip capillary electrophoresis for end-point analysis enables accurate sizing of DNA fragments and identification of fish species at low concentrations.

References

1. Lin, W. F., & Hwang, D. F. (2007). Application of PCR-RFLP analysis on species identification of canned tuna. Food Control, 18(5), 1050–1057. doi: 10.1016/j.foodcont.2006.07.001

2. Chen, C. H., Hsieh, C. H., & Hwang, D. F. (2013). PCR-RFLP analysis using capillary electrophoresis for species identification of Cyprinidae-related products. Food Control, 32(1), 1–8. doi: 10.1016/j.foodcont.2013.01.001
3. Garrett, S., & Dooley, J. (n.d.). Determination of PCR-RFLP Profiles for Fish Species Using the Agilent 2100 Bioanalyzer. Retrieved from https://www.chem-agilent.com/pdf/5989-2982EN.pdf

References

1. Bal, D.V. and K.V. Rao , 1990. Marine fisheries of India. Tata McGraw-Hill Publishing Company Limited, New Delhi. p. 472.
2. Biswas, S.P. (2002). Fundamentals of Ichthyology Narendra Publishing House, Delhi. 392 pp.
3. Day, F. 1878 The fishes of India, being a natural history of the fishes known to inhabit the seas and freshwaters of India, Burma and Ceylon, Part 4: i-xx + 553-779 p. + pls. 139 - 195.
4. Eschmayer W. (Ed) 1998. The Catalog of Fishes. San Francisco: California Academy of Sciences.
5. Fischer, W. and G. Bianchi (Eds.) 1984. FAO species identification sheets for fishery purposes. Western Indian Ocean (Fishing Area 51). Prepared and printed with the support of the Danish International Development Agency (DANIDA). Rome, Food and Agricultural Organization of the United Nations, Vols. 1-6.
6. Ian, S.R., Munro, 1982. The marine and freshwater fishes of Ceylon. Narendra Publishing House, New Delhi. pp. 1-351.
7. Jayaram, K.C. (1999): The freshwater fishes of the Indian region. Narendra Publishing House, Delhi. 551 pp.
8. Jayaraman, K.C. (2002). Fundamentals of Fish Taxonomy. Narendra Publishing House. Delhi, 174. pp.
9. Nelson, J.S., 1976. Fishes of the world. A Wiley - Inter science publication. John Wiley & Sons (N.Y) pp. ix + 1- 416.

10. Ramaiyan, V. and R. Sivakumar, 1991. Sharks, skates and rays (an aid to the identification of sharks, skates and rays of Parangipettai Coast). Centre of Advanced Study in Marine Biology (Annamalai University), Parangipettai, India. 55 pp.
11. Talwar, P.K. and A. G. Jhingran 1991 Inland Fishes of India and adjacent countries. Vol.1, IBH Publishing, New Delhi.
12. Venkatarmani, V.K., P. Jawahar, R. Santhanam and T. Vaitheeswaran, 2004. Marine Ornamental Fishes of Gulf of Mannar. Monograph (ICAR/NATP/CGP), FC & RI, TANUVAS. 176 pp.
13. Venkataramanujam, K., V.K. Venkatramani and N. Ramanathan ,1996. A manual of finfish Taxonomy. Fisheries College and Research Institute. 80 pp.
14. Fish base, 2023. Retrieved from website.
15. Worms, 2023. Retrieved from website.
16. R. Durairaja, S. Aruna and S. Felix, 2016. Manual on Taxonomy of Fin Fish Fisheries College and Research Institute TAMIL NADU FISHERIES UNIVERSITY Ponneri, Thiruvallur District.
17. FAO Fish Identissfication manuals (Western and Eastern Indian Oceans)
18. "Textbook of Fish Biology and Fisheries" by S.S. Khanna and H.R. Singh.
19. K. K. Joshi and K.M. Sreekumar, 2015. Basics of sample collection, preservation and species identification of finfish, Summer School on Recent Advances in Marine Biodiversity Conservation and Management.
20. Hubbs, C. L. and Lagler, K. F. 1941. Guide to the fishes of the great lakes and tributary waters. Bull. Cranbrook Inst. Sci., No. 18: 1-100, Figs. 1-118.
21. Jones, R. J. (2004). A Guide to Wax Preserving Wet Specimens. Journal of Aquatic Preservation and Conservation, 12(3), 158-167.
22. D.J. McClements (2005). Food Emulsions: Principles, Practice and Techniques. 2nd Edition, CRC Press. Boca Raton, FL.
23. D.J. McClements (2007). Understanding and Controlling the Microstructure of Complex Foods. Woodhead Publishing, Cambridge, UK.
24. D.J. McClements & E.A. Decker (2009). Designing Functional Foods. Woodhead Publishing, Cambridge, UK.

9 789359 197128

Printed at Repro India Ltd.